The Works of Henri Poincare
By Henri Poincare

The Measure of Time (1898)

by Henri Poincaré, translated by George Bruce Halsted

THE MEASURE OF TIME

I

So long as we do not go outside the domain of consciousness, the notion of time is relatively clear. Not only do we distinguish without difficulty present sensation from the remembrance of past sensations or the anticipation of future sensations, but we know perfectly well what we mean when we say that of two conscious phenomena which we remember, one was anterior to the other; or that, of two foreseen conscious phenomena, one will be anterior to the other.

When we say that two conscious facts are simultaneous, we mean that they profoundly interpenetrate, so that analysis can not separate them without mutilating them.

The order in which we arrange conscious phenomena does not admit of any arbitrariness. It is imposed upon us and of it we can change nothing.

I have only a single observation to add. For an aggregate of sensations to have become a remembrance capable of classification in time, it must have ceased to be actual, we must have lost the sense of its infinite complexity, otherwise it would have remained present. It must, so to speak, have crystallized around a center of associations of ideas which will be a sort of label. It is only when they thus have lost all life that we can classify our memories in time as a botanist arranges dried flowers in his herbarium.

But these labels can only be finite in number. On that score, psychologic time should be discontinuous. Whence comes the feeling that between any two instants there are others? We arrange our recollections in time, but we know that there remain empty compartments. How could that be, if time were not a form pre-existent in our minds? How could we know there were empty compartments, if these compartments were revealed to us only by their content?

II

But that is not all; into this form we wish to put not only the phenomena of our own consciousness, but those of which other consciousnesses are the theater. But more, we wish to put there physical facts, these I know not what with which we people space and which no consciousness sees directly. This is necessary because without it science could not exist. In a word, psychologic time is given to us and must needs create scientific and physical time. There the difficulty begins, or rather the difficulties, for there are two.

Think of two consciousnesses, which are like two worlds impenetrable one to the other. By what right do we strive to put them into the same mold, to measure them by the same standard? Is it not as if one strove to measure length with a gram or weight with a meter? And besides, why do we speak of measuring? We know perhaps that some fact is anterior to some other, but not *by how much* it is anterior.

Therefore two difficulties: (1) Can we transform psychologic time, which is qualitative, into a quantitative time? (2) Can we reduce to one and the same measure facts which transpire in different worlds?

III

The first difficulty has long been noticed; it has been the subject of long discussions and one may say the question is settled. *We have not a direct intuition of the equality of two intervals of time.* The persons who believe they possess this intuition are dupes of an illusion. When I say, from noon to one the same time passes as from two to three, what meaning has this affirmation?

The least reflection shows that by itself it has none at all. It will only have that which I choose to give it, by a definition which will certainly possess a certain degree of arbitrariness. Psychologists could have done without this definition; physicists and astronomers could not; let us see how they have managed.

To measure time they use the pendulum and they suppose by definition that all the beats of this pendulum are of equal duration. But this is only a first approximation; the temperature, the resistance of the air, the barometric pressure, make the pace of the pendulum vary. If we could escape these sources of error, we should obtain a much closer approximation, but it would still be only an approximation. New causes, hitherto neglected, electric, magnetic or others, would introduce minute perturbations.

In fact, the best chronometers must be corrected from time to time, and the corrections are made by the aid of astronomic observations; arrangements are made so that the sidereal clock marks the same hour when the same star passes the meridian. In other words, it is the sidereal day, that is, the duration of the rotation of the earth, which is the constant unit of time. It is supposed, by a new definition substituted for that based on the beats of the pendulum, that two complete rotations of the earth about its axis have the same duration.

However, the astronomers are still not content with this definition. Many of them think that the tides act as a check on our globe, and that the rotation of the earth is becoming slower and slower. Thus would be explained the apparent acceleration of the motion of the moon, which would seem to be going more rapidly than theory permits because our watch, which is the earth, is going slow.

IV

All this is unimportant, one will say; doubtless our instruments of measurement are imperfect, but it suffices that we can conceive a perfect instrument. This ideal can not be reached, but it is enough to have conceived it and so to have put rigor into the definition of the unit of time.

The trouble is that there is no rigor in the definition. When we use the pendulum to measure time, what postulate do we implicitly admit? *It is that the duration of two identical phenomena is the same*; or, if you prefer, that the same causes take the same time to produce the same effects.

And at first blush, this is a good definition of the equality of two durations. But take care. Is it impossible that experiment may some day contradict our postulate?

Let me explain myself. I suppose that at a certain place in the world the phenomenon α happens, causing as consequence at the end of a certain time the effect α'. At another place in the world very far away from the first, happens the phenomenon β, which causes as consequence the effect β'. The phenomena α and β are simultaneous, as are also the effects α' and β'.

Later, the phenomenon α is reproduced under approximately the same conditions as before, and *simultaneously* the phenomenon β is also reproduced at a very distant place in the world and almost under the same circumstances. The effects α' and β' also take place. Let us suppose that the effect α' happens perceptibly before the effect β'.

If experience made us witness such a sight, our postulate would be contradicted. For experience would tell us that the first duration $\alpha\alpha'$ is equal to the first duration $\beta\beta'$ and that the second duration $\alpha\alpha'$ is less than the second duration $\beta\beta'$. On the other hand, our postulate would require that the two durations $\alpha\alpha'$ should be equal to each other, as likewise the two durations $\beta\beta'$. The equality and the inequality deduced from experience would be incompatible with the two equalities deduced from the postulate.

Now can we affirm that the hypotheses I have just made are absurd? They are in no wise contrary to the principle of contradiction. Doubtless they could not happen without the principle of sufficient reason seeming violated. But to justify a definition so fundamental I should prefer some other guarantee.

V

But that is not all. In physical reality one cause does not produce a given effect, but a multitude of distinct causes contribute to produce it, without our having any means of discriminating the part of each of them.

Physicists seek to make this distinction; but they make it only approximately, and, however they progress, they never will make it except approximately. It is approximately true that the motion of the pendulum is due solely to the earth's attraction; but in all rigor every attraction, even of Sirius, acts on the pendulum.

Under these conditions, it is clear that the causes which have produced a certain effect will never be reproduced except approximately. Then we should modify our postulate and our definition. Instead of saying: 'The same causes take the same time to produce the same effects,' we should say : 'Causes almost identical take almost the same time to produce almost the same effects.'

Our definition therefore is no longer anything but approximate. Besides, as M. Calinon very justly remarks in a recent memoir:[1]

One of the circumstances of any phenomenon is the velocity of the earth's rotation; if this velocity of rotation varies, it constitutes in the reproduction of this phenomenon a circumstance which no longer remains the same. But to suppose this velocity of rotation constant is to suppose that we know how to measure time.

Our definition is therefore not yet satisfactory; it is certainly not that which the astronomers of whom I spoke above implicitly adopt, when they affirm that the terrestrial rotation is slowing down.

What meaning according to them has this affirmation? We can only understand it by analyzing the proofs they give of their proposition. They say first that the friction of the tides producing heat must destroy *vis viva*. They invoke therefore the principle of *vis viva*, or of the conservation of energy.

They say next that the secular acceleration of the moon, calculated according to Newton's law, would be less than that deduced from observations unless the correction relative to the slowing down of the terrestrial rotation were made. They invoke therefore Newton's law. In other words, they define duration in the following way: time should be so defined that Newton's law and that of *vis viva* may be verified. Newton's law is an experimental truth; as such it is only approximate, which shows that we still have only a definition by approximation.

If now it be supposed that another way of measuring time is adopted, the experiments on which Newton's law is founded would none the less have the same meaning. Only the enunciation of the law would be different, because it would be translated into another language; it would evidently be much less simple. So that the definition implicitly adopted by the astronomers may be summed up thus: Time should be so defined that the equations of mechanics may be as simple as possible. In other words, there is not one way of measuring time more true than another; that which is generally adopted is only more *convenient*. Of two watches, we have no right to say that the one goes true, the other wrong; we can only say that it is advantageous to conform to the indications of the first.

The difficulty which has just occupied us has been, as I have said, often pointed out; among the most recent works in which it is considered, I may mention, besides M. Calinon's little book, the treatise on mechanics of Andrade.

VI

The second difficulty has up to the present attracted much less attention; yet it is altogether analogous to the preceding; and even, logically, I should have spoken of it first.

Two psychological phenomena happen in two different consciousnesses; when I say they are simultaneous, what do I mean? When I say that a physical phenomenon, which happens outside of every consciousness, is before or after a psychological phenomenon, what do I mean?

In 1572, Tycho Brahe noticed in the heavens a new star. An immense conflagration had happened in some far distant heavenly body; but it had happened long before; at least two hundred years were necessary for the light from that star to reach our earth. This conflagration therefore happened before the discovery of America. Well, when I say that; when, considering this gigantic phenomenon, which perhaps had no witness, since the satellites of that star were perhaps uninhabited, I say this phenomenon is anterior to the formation of the visual image of the isle of Española in the consciousness of Christopher Columbus, what do I mean?

A little reflection is sufficient to understand that all these affirmations have by themselves no meaning. They can have one only as the outcome of a convention.

VII

We should first ask ourselves how one could have had the idea of putting into the same frame so many worlds impenetrable to one another. We should like to represent to ourselves the external universe, and only by so doing could we feel that we understood it. We know we never can attain this representation: our weakness is too great. But at least we desire the ability to conceive an infinite intelligence for which this representation could be possible, a sort of great consciousness which should see all, and which should classify all *in its time*, as we classify, *in our time*, the little we see.

This hypothesis is indeed crude and incomplete, because this supreme intelligence would be only a demigod; infinite in one sense, it would be limited in another, since it would have only an imperfect recollection of the past; and it could have no other, since otherwise all recollections would be equally present to it and for it there would be no time. And yet when we speak of time, for all which happens outside of us, do we not unconsciously adopt this hypothesis; do we not put ourselves in the place of this imperfect god; and do not even the atheists put themselves in the place where god would be if he existed?

What I have just said shows us, perhaps, why we have tried to put all physical phenomena into the same frame. But that can not pass for a definition of simultaneity, since this hypothetical intelligence, even if it existed, would be for us impenetrable. It is therefore necessary to seek something else.

VIII

The ordinary definitions which are proper for psychologic time would suffice us no more. Two simultaneous psychologic facts are so closely bound together that analysis can not separate without mutilating them. Is it the same with two physical facts? Is not my present nearer my past of yesterday than the present of Sirius?

It has also been said that two facts should be regarded as simultaneous when the order of their succession may be inverted at will. It is evident that this definition would not suit two physical facts which happen far from one another, and that, in what concerns them, we no longer even understand what this reversibility would be; besides, succession itself must first be defined.

IX

Let us then seek to give an account of what is understood by simultaneity or antecedence, and for this let us analyze some examples.

I write a letter; it is afterward read by the friend to whom I have addressed it. There are two facts which have had for their theater two different consciousnesses. In writing this letter I have had the visual image of it, and my friend has had in his turn this same visual image in reading the letter. Though these two facts happen in impenetrable worlds, I do not hesitate to regard the first as anterior to the second, because I believe it is its cause.

I hear thunder, and I conclude there has been an electric discharge; I do not hesitate to consider the physical phenomenon as anterior to the auditory image perceived in my consciousness, because I believe it is its cause.

Behold then the rule we follow, and the only one we can follow: when a phenomenon appears to us as the cause of another, we regard it as anterior. It is therefore by cause that we define time; but most often, when two facts appear to us bound by a constant relation, how do we recognize which is the cause and which the effect? We assume that the anterior fact, the antecedent, is the cause of the other, of the consequent. It is then by time that we define cause. How save ourselves from this *petitio principii*?

We say now *post hoc, ergo propter hoc*; now *propter hoc, ergo post hoc*; shall we escape from this vicious circle?

X

Let us see, not how we succeed in escaping, for we do not completely succeed, but how we try to escape.

I execute a voluntary act A and I feel afterward a sensation D, which I regard as a consequence of the act A; on the other hand, for whatever reason, I infer that this consequence is not immediate, but that outside my consciousness two facts B and C, which I have not witnessed, have happened, and in such a way that B is the effect of A, that C is the effect of B, and D of C.

But why? If I think I have reason to regard the four facts A, B, C, D, as bound to one another by a causal connection, why range them in the causal order $ABCD$, and at the same time in the chronologic order $ABCD$, rather than in any other order?

I clearly see that in the act A I have the feeling of having been active, while in undergoing the sensation D I have that of having been passive. This is why I regard A as the initial cause and D as the ultimate effect; this is why I put A at the beginning of the chain and D at the end; but why put B before C rather than C before B?

If this question is put, the reply ordinarily is: we know that it is B which is the cause of C because we always see B happen before C. These two phenomena, when witnessed, happen in a certain order; when analogous phenomena happen without witness, there is no reason to invert this order.

Doubtless, but take care; we never know directly the physical phenomena B and C. What we know are sensations B' and C' produced respectively by B and C. Our consciousness tells us immediately that B' precedes C' and we suppose that B and C succeed one another in the same order.

This rule appears in fact very natural, and yet we are often led to depart from it. We hear the sound of the thunder only some seconds after the electric discharge of the cloud. Of two flashes of lightning, the one distant, the other near, can not the first be anterior to the second, even though the sound of the second comes to us before that of the first?

XI

Another difficulty; have we really the right to speak of the cause of a phenomenon? If all the parts of the universe are interchained in a certain measure, any one phenomenon will not be the effect of a single cause, but the resultant of causes infinitely numerous; it is, one often says, the consequence of the state of the universe a moment before. How enunciate rules applicable to circumstances so complex? And yet it is only thus that these rules can be general and rigorous.

Not to lose ourselves in this infinite complexity, let us make a simpler hypothesis. Consider three stars, for example, the sun, Jupiter and Saturn; but, for greater simplicity, regard them as reduced to material points and isolated from the rest of the world. The positions and the velocities of three bodies at a given instant suffice to determine their positions and velocities at the following instant, and consequently at any instant. Their positions at the instant t determine their positions at the instant $t + h$ as well as their positions at the instant $t - h$.

Even more; the position of Jupiter at the instant t, together with that of Saturn at the instant $t + a$, determines the position of Jupiter at any instant and that of Saturn at any instant

The aggregate of positions occupied by Jupiter at the instant $t + e$ and Saturn at the instant $t + a + e$ is bound to the aggregate of positions occupied by Jupiter at the instant t and Saturn at the instant $t + a$, by laws as precise as that of Newton, though more complicated. Then why not regard one of these aggregates as the cause of the other, which would lead to considering as simultaneous the instant t of Jupiter and the instant $t + a$ of Saturn?

In answer there can only be reasons, very strong, it is true, of convenience and simplicity.

XII

But let us pass to examples less artificial; to understand the definition implicitly supposed by the savants, let us watch them at work and look for the rules by which they investigate simultaneity.

I will take two simple examples, the measurement of the velocity of light and the determination of longitude.

When an astronomer tells me that some stellar phenomenon, which his telescope reveals to him at this moment, happened, nevertheless, fifty years ago, I seek his meaning, and to that end I shall ask him first how he knows it, that is, how he has measured the velocity of light.

He has begun by *supposing* that light has a constant velocity, and in particular that its velocity is the same in all directions. That is a postulate without which no measurement of this velocity could be attempted. This postulate could never be verified directly by experiment; it might be contradicted by it if the results of different measurements were not concordant. We should think ourselves fortunate that this contradiction has not happened and that the slight discordances which may happen can be readily explained.

The postulate, at all events, resembling the principle of sufficient reason, has been accepted by everybody; what I wish to emphasize is that it furnishes us with a new rule for the investigation of simultaneity, entirely different from that which we have enunciated above.

This postulate assumed, let us see how the velocity of light has been measured. You know that Roemer used eclipses of the satellites of Jupiter, and sought how much the event fell behind its prediction. But how is this prediction made? It is by the aid of astronomic laws; for instance Newton's law.

Could not the observed facts be just as well explained if we attributed to the velocity of light a little different value from that adopted, and supposed Newton's law only approximate? Only this would lead to replacing Newton's law by another more complicated. So for the velocity of light a value is adopted, such that the astronomic laws compatible with this value may be as simple as possible. When navigators or geographers determine a longitude, they have to solve just the problem we are discussing; they must, without being at Paris, calculate Paris time. How do they accomplish it? They carry a chronometer set for Paris. The qualitative problem of simultaneity is made to depend upon the quantitative problem of the measurement of time. I need not take up the difficulties relative to this latter problem, since above I have emphasized them at length.

Or else they observe an astronomic phenomenon, such as an eclipse of the moon, and they suppose that this phenomenon is perceived simultaneously from all points of the earth. That is not altogether true, since the propagation of light is not instantaneous; if absolute exactitude were desired, there would be a correction to make according to a complicated rule.

Or else finally they use the telegraph. It is clear first that the reception of the signal at Berlin, for instance, is after the sending of this same signal from Paris. This is the rule of cause and effect analyzed above. But how much after? In general, the duration of the transmission is neglected and the two events are regarded as simultaneous. But, to be rigorous, a little correction would still have to be made by a complicated calculation; in practise it is not made, because it would be well within the errors of observation; its theoretic necessity is none the less from our point of view, which is that of a rigorous definition. From this discussion, I wish to emphasize two things: (1) The rules applied are exceedingly various. (2) It is difficult to separate the qualitative problem of simultaneity from the quantitative problem of the measurement of time; no matter whether a chronometer is used, or whether account must be taken of a velocity of transmission, as that of light, because such a velocity could not be measured without *measuring* a time.

XIII

To conclude: We have not a direct intuition of simultaneity, nor of the equality of two durations. If we think we have this intuition, this is an illusion. We replace it by the aid of certain rules which we apply almost always without taking count of them.

But what is the nature of these rules? No general rule, no rigorous rule; a multitude of little rules applicable to each particular case.

These rules are not imposed upon us and we might amuse ourselves in inventing others; but they could not be cast aside without greatly complicating the enunciation of the laws of physics, mechanics and astronomy.

We therefore choose these rules, not because they are true, but because they are the most convenient, and we may recapitulate them as follows: "The simultaneity of two events, or the order of their succession, the equality of two durations, are to be so defined that the enunciation of the natural laws may be as simple as possible. In other words, all these rules, all these definitions are only the fruit of an unconscious opportunism."

1. *Etude sur les diverses grandeurs*, Paris, Gauthier-Villars, 1897.

SCIENCE AND HYPOTHESIS

BY

H. POINCARÉ,

MEMBER OF THE INSTITUTE OF FRANCE.

WITH A PREFACE BY

J. LARMOR, D.Sc., SEC. R.S.,

Lucasian Professor of Mathematics in the University of Cambridge,

London and Newcastle-on-Tyne:

THE WALTER SCOTT PUBLISHING CO., LTD.

NEW YORK: 3 EAST 14TH STREET.

1905.

TRANSLATOR'S NOTE

———————————

THE translator wishes to express his indebtedness to Professor Larmor, for kindly consenting to introduce the author of *Science and Hypothesis* to English readers; to Dr. F. S. Macaulay and Mr. C. S. Jackson, M.A., who have read the whole of the proofs and have greatly helped by suggestions; also to Professor G. H. Bryan, F.R.S., who has read the proofs of Chapter VIII., and whose criticisms have been most valuable.

W. J. G.

February 1905.

INTRODUCTION.

IT is to be hoped that, as a consequence of the present active scrutiny of our educational aims and methods, and of the resulting encouragement of the study of modern languages, we shall not remain, as a nation, so much isolated from ideas and tendencies in continental thought and literature as we have been in the past. As things are, however, the translation of this book is doubtless required; at any rate, it brings vividly before us an instructive point of view. Though some of M. Poincaré's chapters have been collected from well-known treatises written several years ago, and indeed are sometimes in detail not quite up to date, besides occasionally suggesting the suspicion that his views may possibly have been modified in the interval, yet their publication in a compact form has excited a warm welcome in this country.

It must be confessed that the English language hardly lends itself as a perfect medium for the rendering of the delicate shades of suggestion and allusion characteristic of M. Poincaré's play around his subject; notwithstanding the excellence of the translation, loss in this respect is inevitable.

There has been of late a growing trend of opinion, prompted in part by general philosophical views, in the direction that the theoretical constructions of physical science are largely factitious, that instead of presenting a valid image of the relations of things on which further progress can be based, they are still little better than a mirage. The best method of abating this scepticism is to become acquainted with the real scope and modes of application of conceptions which, in the popular language of superficial exposition — and even in the unguarded and playful paradox of their authors, intended only for the instructed eye — often look bizarre enough. But much advantage will accrue if men of science become their own epistemologists, and show to the world by critical exposition in non-technical terms of the results and methods of their constructive work, that more than mere instinct is involved in it: the community has indeed a right to expect as much as this. It would be hard to find any one better qualified for this kind of exposition, either from the profundity of his own mathematical achievements, or from the extent and freshness of his interest in the theories of physical science, than the author of this book. If an appreciation might be ventured on as regards the later chapters, they are, perhaps, intended to present the stern logical analyst quizzing the cultivator of physical ideas as to what he is driving at, and whither he expects to go, rather than any responsible attempt towards a settled confession of faith. Thus, when M. Poincaré allows himself for a moment to indulge in a process of evaporation of the Principle of Energy, he is content to sum up: "Eh bien, quelles que soient les notions nouvelles que les expériences futures nous donneront sur le monde, nous sommes sûrs d'avance qu'il y aura quelque chose qui demeurera constant et que nous pourrons appeler *énergie*" (p. 166), and to leave the matter there for his readers to think it out. Though hardly necessary in the original French, it may not now be superfluous to point out that independent reflection and criticism on the part of the reader are tacitly implied here as else where.

An interesting passage is the one devoted to Maxwell's theory of the functions of the æther, and the comparison of the close-knit theories of the classical French mathematical physicists with the somewhat loosely-connected *corpus* of ideas by which Maxwell, the interpreter and successor of Faraday, has (posthumously) recast the whole face of physical science. How many times has that theory been re-written since Maxwell's day? and yet how little has it been altered in essence, except by further developments in the problem of moving bodies, from the form in which he left it! If, as M. Poincaré remarks, the French instinct for precision and lucid demonstration sometimes finds itself ill at ease with physical theories of the British school, he as readily admits (pp. 223, 224), and indeed fully appreciates, the advantages on the other side. Our own mental philosophers have been shocked at the point of view indicated by the proposition hazarded by Laplace, that a sufficiently developed intelligence, if it were made acquainted with the positions and motions of the atoms at any instant, could predict all future history: no amount of demur suffices sometimes to persuade them that this is not a conception universally entertained in physical science. It was not so even in Laplace's own day. From the point of view of the study of the evolution of the sciences, there are few episodes more instructive than the collision between Laplace and Young with regard to the theory of capillarity. The precise and intricate mathematical analysis of Laplace, starting from fixed preconceptions regarding atomic forces which were to remain intact throughout the logical development of the argument, came into contrast with the tentative, mobile intuitions of Young; yet the latter was able to grasp, by sheer direct mental force, the fruitful though partial analogies of this recondite class of phenomena with more familiar operations of nature, and to form a direct picture of the way things interacted, such as could only have been illustrated, quite possibly damaged or obliterated, by premature effort to translate it into elaborate analytical formulas. The *aperçus* of Young were apparently devoid of all cogency to Laplace; while Young expressed, doubtless in too extreme a way, his sense of the inanity of the array of mathematical logic of his rival. The subsequent history involved the Nemesis that the fabric of Laplace was taken down and reconstructed in the next generation by Poisson; while the modern cultivator of the subject turns, at any rate in England, to neither of those expositions for illumination, but rather finds in the partial and succinct indications of Young the best starting-point for further effort.

It seems, however, hard to accept entirely the distinction suggested (p. 213) between the methods of cultivating theoretical physics in the two countries. To mention only two transcendent names which stand at the very front of two of the greatest developments of physical science of the last century, Carnot and Fresnel, their procedure was certainly not on the lines thus described. Possibly it is not devoid of significance that each of them attained his first effective recognition from the British school.

It may, in fact, be maintained that the part played by mechanical and such-like theories — analogies if you will — is an essential one. The reader of this book will appreciate that the human mind has need of many instruments of comparison and discovery besides the unrelenting logic of the infinitesimal calculus. The dynamical basis which underlies the objects of our most frequent experience has now been systematised into a great calculus of exact thought, and traces of new real relationships may come out more vividly when considered in terms of our familiar acquaintance with dynamical systems than when formulated under the paler shadow of more analytical abstractions. It is even possible for a constructive physicist to conduct his mental operations entirely by dynamical images, though Helmholtz, as well as our author, seems to class a predilection in this direction as a British trait. A time arrives when, as in other subjects, ideas have crystallised out into distinctness; their exact verification and development then becomes a problem in mathematical physics. But whether the mechanical analogies still survive, or new terms are now introduced devoid of all naive mechanical bias, it matters essentially little. The precise de termination of the relations of things in the rational scheme of nature in which we find ourselves is the fundamental task, and for its fulfilment in any direction advantage has to be taken of our knowledge, even when only partial, of new aspects and types of relationship which may have become familiar perhaps in quite different fields. Nor can it be forgotten that the most fruitful and fundamental conceptions of abstract pure mathematics itself have often been suggested from these mechanical ideas of flux and force, where the play of intuition is our most powerful guide. The study of the historical evolution of physical theories is essential to the complete understanding of their import. It is in the mental workshop of a Fresnel, a Kelvin, or a Helmholtz, that profound ideas of the deep things of Nature are struck out and assume form; when pondered over and paraphrased by philosophers we see them react on the conduct of life: it is the business of criticism to polish them gradually to the common measure of human understanding. Oppressed though we are with the necessity of being specialists, if we are to know anything thoroughly in these days of accumulated details, we may at any rate profitably study the historical evolution of knowledge over a field wider than our own.

The aspect of the subject which has here been dwelt on is that scientific progress, considered historically, is not a strictly logical process, and does not proceed by syllogisms. New ideas emerge dimly into intuition, come into consciousness from nobody knows where, and become the material on which the mind operates, forging them gradually into consistent doctrine, which can be welded on to existing domains of know ledge. But this process is never complete: a crude connection can always be pointed to by a logician as an indication of the imperfection of human constructions.

If intuition plays a part which is so important, it is surely necessary that we should possess a firm grasp of its limitations. In M. Poincaré's earlier chapters the reader can gain very pleasantly a vivid idea of the various and highly complicated ways of docketing our perceptions of the relations of external things, all equally valid, that were open to the human race to develop. Strange to say, they never tried any of them; and, satisfied with the very remarkable practical fitness of the scheme of geometry and dynamics that came naturally to hand, did not consciously trouble themselves about the possible existence of others until recently. Still more recently has it been found that the good Bishop Berkeley's logical jibes against the Newtonian ideas of fluxions and limiting ratios cannot be adequately appeased in the rigorous mathematical conscience, until our apparent continuities are resolved mentally into discrete aggregates which we only partially apprehend. The irresistible impulse to atomize everything thus proves to be not merely a disease of the physicist; a deeper origin, in the nature of knowledge itself, is suggested.

Everywhere want of absolute, exact adaptation can be detected, if pains are taken, between the various constructions that result from our mental activity and the impressions which give rise to them. The bluntness of our unaided sensual perceptions, which are the source in part of the intuitions of the race, is well brought out in this connection by M. Poincaré. Is there real contradiction? Harmony usually proves to be re covered by shifting our attitude to the phenomena. All experience leads us to interpret the totality of things as a consistent cosmos — undergoing evolution, the naturalists will say — in the large-scale workings of which we are interested spectators and explorers, while of the inner relations and ramifications we only apprehend dim glimpses. When our formulation of experience is imperfect or even paradoxical, we learn to attribute the fault to our point of view, and to expect that future adaptation will put it right. But Truth resides in a deep well, and we shall never get to the bottom. Only, while deriving enjoyment and insight from M. Poincaré's Socratic exposition of the limitations of the human outlook on the universe, let us beware of counting limitation as imperfection, and drifting into an inadequate conception of the wonderful fabric of human knowledge.

J. LARMOR.

AUTHOR'S PREFACE.

To the superficial observer scientific truth is unassailable, the logic of science is infallible; and if scientific men sometimes make mistakes, it is because they have not understood the rules of the game. Mathematical truths are derived from a few self-evident propositions, by a chain of flawless reasonings; they are imposed not only on us, but on Nature itself. By them the Creator is fettered, as it were, and His choice is limited to a relatively small number of solutions. A few experiments, therefore, will be sufficient to enable us to determine what choice He has made. From each experiment a number of consequences will follow by a series of mathematical deductions, and in this way each of them will reveal to us a corner of the universe. This, to the minds of most people, and to students who are getting their first ideas of physics, is the origin of certainty in science. This is what they take to be the role of experiment and mathematics. And thus, too, it was understood a hundred years ago by many men of science who dreamed of constructing the world with the aid of the smallest possible amount of material borrowed from experiment.

But upon more mature reflection the position held by hypothesis was seen; it was recognised that it is as necessary to the experimenter as it is to the mathematician. And then the doubt arose if all these constructions are built on solid foundations. The conclusion was drawn that a breath would bring them to the ground. This sceptical attitude does not escape the charge of superficiality. To doubt everything or to believe everything are two equally convenient solutions; both dispense with the necessity of reflection.

Instead of a summary condemnation we should examine with the utmost care the role of hypothesis; we shall then recognise not only that it is necessary, but that in most cases it is legitimate. We shall also see that there are several kinds of hypotheses; that some are verifiable, and when once confirmed by experiment become truths of great fertility; that others may be useful to us in fixing our ideas; and finally, that others are hypotheses only in appearance, and reduce to definitions or to conventions in disguise. The latter are to be met with especially in mathematics and in the sciences to which it is applied. From them, indeed, the sciences derive their rigour; such conventions are the result of the unrestricted activity of the mind, which in this domain recognises no obstacle. For here the mind may affirm because it lays down its own laws; but let us clearly understand that while these laws are imposed on *our* science, which otherwise could not exist, they are not imposed on Nature. Are they then arbitrary? No; for if they were, they would not be fertile. Experience leaves us our freedom of choice, but it guides us by helping us to discern the most convenient path to follow. Our laws are therefore like those of an absolute monarch, who is wise and consults his council of state. Some people have been struck by this characteristic of free convention which may be recognised in certain fundamental principles of the sciences. Some have set no limits to their generalisations, and at the same time they have forgotten that there is a difference between liberty and the purely arbitrary. So that they are compelled to end in what is called *nominalism*; they have asked if the *savant* is not the dupe of his own definitions, and if the world he thinks he has discovered is not simply the creation of his own caprice.[1] Under these conditions science would retain its certainty, but would not attain its object, and would become powerless. Now, we daily see what science is doing for us. This could not be unless it taught us something about reality; the aim of science is not things themselves, as the dogmatists in their simplicity imagine, but the relations between things; outside those relations there is no reality knowable.

Such is the conclusion to which we are led; but to reach that conclusion we must pass in review the series of sciences from arithmetic and geometry to mechanics and experimental physics. What is the nature of mathematical reasoning? Is it really deductive, as is commonly supposed? Careful analysis shows us that it is nothing of the kind; that it participates to some extent in the nature of inductive reasoning, and for that reason it is fruitful. But none the less does it retain its character of absolute rigour; and this is what must first be shown.

When we know more of this instrument which is placed in the hands of the investigator by mathematics, we have then to analyse another fundamental idea, that of mathematical magnitude. Do we find it in nature, or have we our selves introduced it? And if the latter be the case, are we not running a risk of coming to incorrect conclusions all round? Comparing the rough data of our senses with that extremely complex and subtle conception which mathematicians call magnitude, we are compelled to recognise a divergence. The framework into which we wish to make everything fit is one of our own construction; but we did not construct it at random, we constructed it by measurement so to speak; and that is why we can fit the facts into it without altering their essential qualities.

Space is another framework which we impose on the world. Whence are the first principles of geometry derived? Are they imposed on us by logic? Lobatschewsky, by inventing non-Euclidean geometries, has shown that this is not the case. Is space revealed to us by our senses? No; for the space revealed to us by our senses is absolutely different from the space of geometry. Is geometry derived from experience? Careful discussion will give the answer — no! We therefore conclude that the principles of geometry are only conventions; but these conventions are not arbitrary, and if transported into another world (which I shall call the non-Euclidean world, and which I shall endeavour to describe), we shall find ourselves compelled to adopt more of them.

In mechanics we shall be led to analogous conclusions, and we shall see that the principles of this science, although more directly based on experience, still share the conventional character of the geometrical postulates. So far, nominalism triumphs; but we now come to the physical sciences, properly so called, and here the scene changes. We meet with hypotheses of another kind, and we fully grasp how fruitful they are. No doubt at the outset theories seem unsound, and the history of science shows us how ephemeral they are; but they do not entirely perish, and of each of them some traces still remain. It is these traces which we must try to discover, because in them and in them alone is the true reality.

The method of the physical sciences is based upon the induction which leads us to expect the recurrence of a phenomenon when the circumstances which give rise to it are repeated. If all the circumstances could be simultaneously reproduced, this principle could be fearlessly applied; but this never happens; some of the circumstances will always be missing. Are we absolutely certain that they are unimportant? Evidently not! It may be probable, but it cannot be rigorously certain. Hence the importance of the role that is played in the physical sciences by the law of probability. The calculus of probabilities is there fore not merely a recreation, or a guide to the baccarat player; and we must thoroughly examine the principles on which it is based. In this connection I have but very incomplete results to lay before the reader, for the vague instinct which enables us to determine probability almost defies analysis. After a study of the conditions under which the work of the physicist is carried on, I have thought it best to show him at work. For this purpose I have taken instances from the history of optics and of electricity. We shall thus see how the ideas of Fresnel and

Maxwell took their rise, and what unconscious hypotheses were made by Ampère and the other founders   of   electro-dynamics.

Footnotes

1. Cf. M. le Roy: "Science et Philosophie," *Revue de Métaphysique et de Morale*, 1901.
Part I: Number and Magnitude

SCIENCE AND HYPOTHESIS.

PART I.

NUMBER AND MAGNITUDE.

CHAPTER I.
ON THE NATURE OF MATHEMATICAL REASONING.

I.

THE very possibility of mathematical science seems an insoluble contradiction. If this science is only deductive in appearance, from whence is derived that perfect rigour which is challenged by none? If, on the contrary, all the propositions which it enunciates may be derived in order by the rules of formal logic, how is it that mathematics is not reduced to a gigantic tautology? The syllogism can teach us nothing essentially new, and if everything must spring from the principle of identity, then everything should be capable of being reduced to that principle. Are we then to admit that the enunciations of all the theorems with which so many volumes are filled, are only indirect ways of saying that A is A?

No doubt we may refer back to axioms which are at the source of all these reasonings. If it is felt that they cannot be reduced to the principle of contradiction, if we decline to see in them any more than experimental facts which have no part or lot in mathematical necessity, there is still one resource left to us: we may class them among *à priori* synthetic views. But this is no solution of the difficulty it is merely giving it a name; and even if the nature of the synthetic views had no longer for us any mystery, the contradiction would not have disappeared; it would have only been shirked. Syllogistic reasoning remains incapable of adding anything to the data that are given it; the data are reduced to axioms, and that is all we should find in the conclusions.

No theorem can be new unless a new axiom intervenes in its demonstration; reasoning can only give us immediately evident truths borrowed from direct intuition; it would only be an intermediary parasite. Should we not therefore have reason for asking if the syllogistic apparatus serves only to disguise what we have borrowed?

The contradiction will strike us the more if we open any book on mathematics; on every page the author announces his intention of generalising some proposition already known. Does the mathematical method proceed from the particular to the general, and, if so, how can it be called deductive? Finally, if the science of number were merely analytical, or could be analytically derived from a few synthetic intuitions, it seems that a sufficiently powerful mind could with a single glance perceive all its truths; nay, one might even hope that some day a language would be invented simple enough for these truths to be made evident to any person of ordinary intelligence.

Even if these consequences are challenged, it must be granted that mathematical reasoning has of itself a kind of creative virtue, and is therefore to be distinguished from the syllogism. The difference must be profound. We shall not, for instance, find the key to the mystery in the frequent use of the rule by which the same uniform operation applied to two equal numbers will give identical results. All these modes of reasoning, whether or not reducible to the syllogism, properly so called, retain the analytical character, and *ipso facto*, lose their power.

II.

The argument is an old one. Let us see how Leibnitz tried to show that two and two make four. I assume the number one to be defined, and also the operation $x+1$ — i.e., the adding of unity to a given number x. These definitions, whatever they may be, do not enter into the subsequent reasoning. I next define the numbers 2, 3, 4 by the equalities: —

(1) $1 + 1 = 2$; (2) $2 + 1 = 3$; (3) $3 + 1 = 4$ and in the same way I define the operation $x + 2$ by the relation; (4) $x + 2 = (x + 1) + 1$. Given this, we have:

$$2 + 2 = (2 + 1) + 1; \quad \text{(def. 4)}.$$
$$(2 + 1) + 1 = 3 + 1 \quad \text{(def. 2)}.$$
$$3 + 1 = 4 \quad \text{(def. 3)}.$$
$$\text{whence } 2 + 2 = 4 \quad \text{Q.E.D.}$$

It cannot be denied that this reasoning is purely analytical. But if we ask a mathematician, he will reply: "This is not a demonstration properly so called; it is a verification." We have confined ourselves to bringing together one or other of two purely conventional definitions, and we have verified their identity; nothing new has been learned. *Verification* differs from proof precisely because it is analytical, and because it leads to nothing. It leads to nothing because the conclusion is nothing but the premisses translated into another language. A real proof, on the other hand, is fruitful, because the conclusion is in a sense more general than the premisses. The equality $2 + 2 = 4$ can be verified because it is particular. Each individual enunciation in mathematics may be always verified in the same way. But if mathematics could be reduced to a series of such verifications it would not be a science. A chess-player, for instance, does not create a science by winning a piece. There is no science but the science of the general. It may even be said that the object of the exact sciences is to dispense with these direct verifications.

III.

Let us now see the geometer at work, and try to surprise some of his methods. The task is not without difficulty; it is not enough to open a book at random and to analyse any proof we may come across. First of all, geometry must be excluded, or the question becomes complicated by difficult problems relating to the role of the postulates, the nature and the origin of the idea of space. For analogous reasons we cannot avail ourselves of the infinitesimal calculus. We must seek mathematical thought where it has remained pure — i.e., in Arithmetic. But we still have to choose; in the higher parts of the theory of numbers the primitive mathematical ideas have already undergone so profound an elaboration that it becomes difficult to analyse them.

It is therefore at the beginning of Arithmetic that we must expect to find the explanation we seek; but it happens that it is precisely in the proofs of the most elementary theorems that the authors of classic treatises have displayed the least precision and rigour. We may not impute this to them as a crime; they have obeyed a necessity. Beginners are not prepared for real mathematical rigour; they would see in it nothing but empty, tedious subtleties. It would be waste of time to try to make them more exacting; they have to pass rapidly and without stopping over the road which was trodden slowly by the founders of the science.

Why is so long a preparation necessary to habituate oneself to this perfect rigour, which it would seem should naturally be imposed on all minds? This is a logical and psychological problem which is well worthy of study. But we shall not dwell on it; it is foreign to our subject. All I wish to insist on is, that we shall fail in our purpose unless we reconstruct the proofs of the elementary theorems, and give them, not the rough form in which they are left so as not to weary the beginner, but the form which will satisfy the skilled geometer.

DEFINITION OF ADDITION.

I assume that the operation $x + 1$ has been defined; it consists in adding the number 1 to a given number x. Whatever may be said of this definition, it does not enter into the subsequent reasoning.

We now have to define the operation $x + a$, which consists in adding the number a to any given number x. Suppose that we have defined the operation $x + (a - 1)$; the operation $x + a$ will be defined by the equality: (1) $x + a = [x + (a - 1)] + 1$. We shall know what $x + a$ is when we know what $x + (a - 1)$ is, and as I have assumed that to start with we know what $x + 1$ is, we can define successively and "by recurrence" the operations $x + 2$, $x + 3$, etc. This definition deserves a moment's attention; it is of a particular nature which distinguishes it even at this stage from the purely logical definition; the equality (1), in fact, contains an infinite number of distinct definitions, each having only one meaning when we know the meaning of its predecessor.

PROPERTIES OF ADDITION.

Associative. — I say that $a + (b + c) = (a + b) + c$; in fact, the theorem is true for $c = 1$. It may then be written $a + (b + 1) = (a + b) + 1$; which, remembering the difference of notation, is nothing but the equality (1) by which I have just defined addition. Assume the theorem true for $c = \gamma$, I say that it will be true for $c = \gamma + 1$. Let $(a + b) + \gamma = a + (b + \gamma)$, it follows that $[(a + b) + \gamma] + 1 = [a + (b + \gamma)] + 1$; or by def. (1) — $(a + b) + (\gamma + 1) = a + (b + \gamma + 1) = a + [b + (\gamma + 1)]$, which shows by a series of purely analytical deductions that the theorem is true for $\gamma + 1$. Being true for $c = 1$, we see that it is successively true for $c = 2$, $c = 3$, etc.

Commutative. (1) I say that $a + 1 = 1 + a$. The theorem is evidently true for $a = 1$; we can *verify* by purely analytical reasoning that if it is true for $a = \gamma$ it will be true for $a = \gamma + 1$.[1]Now, it is true for $a = 1$, and therefore is true for $a = 2$, $a = 3$, and so on. This is what is meant by saying that the proof is demonstrated "by recurrence."

(2) I say that $a + b = b + a$. The theorem has just been shown to hold good for $b = 1$, and it may be verified analytically that if it is true for $b = \beta$ it will be true for $b = \beta + 1$. The proposition is thus established by recurrence.

DEFINITION OF MULTIPLICATION.

We shall define multiplication by the equalities: (1) $a \times 1 = a$. (2) $a \times b = [a \times (b - 1)] + a$. Both of these include an infinite number of definitions; having defined $a \times 1$, it enables us to define in succession $a \times 2$, $a \times 3$, and so on.

PROPERTIES OF MULTIPLICATION.

Distributive. — I say that $(a + b) \times c = (a \times c) + (b \times c)$. We can verify analytically that the theorem is true for $c = 1$; then if it is true for $c = \gamma$, it will be true for $c = \gamma + 1$. The proposition is then proved by recurrence.

Commutative. — (1) I say that $a \times 1 = 1 \times a$. The theorem is obvious for $a = 1$. We can verify analytically that if it is true for $a = \alpha$, it will be true for $a = \alpha + 1$.

(2) I say that $a \times b = b \times a$. The theorem has just been proved for $b = 1$. We can verify analytically that if it be true for $b = \beta$ it will be true for $b = \beta + 1$.

IV.

This monotonous series of reasonings may now be laid aside; but their very monotony brings vividly to light the process, which is uniform, and is met again at every step. The process is proof by recurrence. We first show that a theorem is true for $n = 1$; we then show that if it is true for $n - 1$ it is true for n, and we conclude that it is true for all integers. We have now seen how it may be used for the proof of the rules of addition and multiplication — that is to say, for the rules of the algebraical calculus. This calculus is an instrument of transformation which lends itself to many more different combinations than the simple syllogism; but it is still a purely analytical instrument, and is incapable of teaching us anything new. If mathematics had no other instrument, it would immediately be arrested in its development; but it has recourse anew to the same process — *i.e.*, to reasoning by recurrence, and it can continue its forward march. Then if we look carefully, we find this mode of reasoning at every step, either under the simple form which we have just given to it, or under a more or less modified form. It is therefore mathematical reasoning *par excellence*, and we must examine it closer.

V.

The essential characteristic of reasoning by recurrence is that it contains, condensed, so to speak, in a single formula, an infinite number of syllogisms. We shall see this more clearly if we enunciate the syllogisms one after another. They follow one another, if one may use the expression, in a cascade. The following are the hypothetical syllogisms: — The theorem is true of the number 1. Now, if it is true of 1, it is true of 2; therefore it is true of 2. Now, if it is true of 2, it is true of 3; hence it is true of 3, and so on. We see that the conclusion of each syllogism serves as the minor of its successor. Further, the majors of all our syllogisms may be reduced to a single form. If the theorem is true of $n - 1$, it is true of n.

We see, then, that in reasoning by recurrence we confine ourselves to the enunciation of the minor of the first syllogism, and the general formula which contains as particular cases all the majors. This unending series of syllogisms is thus reduced to a phrase of a few lines.

It is now easy to understand why every particular consequence of a theorem may, as I have above explained, be verified by purely analytical processes. If, instead of proving that our theorem is true for all numbers, we only wish to show that it is true for the number 6 for instance, it will be enough to establish the first five syllogisms in our cascade. We shall require 9 if we wish to prove it for the number 10; for a greater number we shall require more still; but however great the number may be we shall always reach it, and the analytical verification will always be possible. But however far we went we should never reach the general theorem applicable to all numbers, which alone is the object of science. To reach it we should require an infinite number of syllogisms, and we should have to cross an abyss which the patience of the analyst, restricted to the resources of formal logic, will never succeed in crossing.

I asked at the outset why we cannot conceive of a mind powerful enough to see at a glance the whole body of mathematical truth. The answer is now easy. A chess-player can combine for four or five moves ahead; but, however extraordinary a player he may be, he cannot prepare for more than a finite number of moves. If he applies his faculties to Arithmetic, he cannot conceive its general truths by direct intuition alone; to prove even the smallest theorem he must use reasoning by recurrence, for that is the only instrument which enables us to pass from the finite to the infinite. This instrument is always useful, for it enables us to leap over as many stages as we wish; it frees us from the necessity of long, tedious, and monotonous verifications which would rapidly become impracticable. Then when we take in hand the general theorem it becomes indispensable, for otherwise we should ever be approaching the analytical verification without ever actually reaching it. In this domain of Arithmetic we may think ourselves very far from the infinitesimal analysis, but the idea of mathematical infinity is already playing a preponderating part, and without it there would be no science at all, because there would be nothing general.

VI.

The views upon which reasoning by recurrence is based may be exhibited in other forms; we may say, for instance, that in any finite collection of different integers there is always one which is smaller than any other. We may readily pass from one enunciation to another, and thus give our selves the illusion of having proved that reasoning by recurrence is legitimate. But we shall always be brought to a full stop — we shall always come to an indemonstrable axiom, which will at bottom be but the proposition we had to prove translated into another language. We cannot therefore escape the conclusion that the rule of reasoning by recurrence is irreducible to the principle of contradiction. Nor can the rule come to us from experiment. Experiment may teach us that the rule is true for the first ten or the first hundred numbers, for instance; it will not bring us to the indefinite series of numbers, but only to a more or less long, but always limited, portion of the series.

Now, if that were all that is in question, the principle of contradiction would be sufficient, it would always enable us to develop as many syllogisms as we wished. It is only when it is a question of a single formula to embrace an infinite number of syllogisms that this principle breaks down, and there, too, experiment is powerless to aid. This rule, inaccessible to analytical proof and to experiment, is the exact type of the *à priori* synthetic intuition. On the other hand, we cannot see in it a convention as in the case of the postulates of geometry.

Why then is this view imposed upon us with such an irresistible weight of evidence? It is because it is only the affirmation of the power of the mind which knows it can conceive of the indefinite repetition of the same act, when the act is once possible. The mind has a direct intuition of this power, and experiment can only be for it an opportunity of using it, and thereby of becoming conscious of it.

But it will be said, if the legitimacy of reasoning by recurrence cannot be established by experiment alone, is it so with experiment aided by induction? We see successively that a theorem is true of the number 1, of the number 2, of the number 3, and so on — the law is manifest, we say, and it is so on the same ground that every physical law is true which is based on a very large but limited number of observations.

It cannot escape our notice that here is a striking analogy with the usual processes of induction. But an essential difference exists. Induction applied to the physical sciences is always uncertain, because it is based on the belief in a general order of the universe, an order which is external to us. Mathematical induction — *i.e.*, proof by recurrence — is, on the contrary, necessarily imposed on us, because it is only the affirmation of a property of the mind itself.

VII.

Mathematicians, as I have said before, always endeavour to generalise the propositions they have obtained. To seek no further example, we have just shown the equality, $a + 1 = 1 + a$, and we then used it to establish the equality, $a + b = b + a$, which is obviously more general. Mathematics may, therefore, like the other sciences, proceed from the particular to the general. This is a fact which might otherwise have appeared incomprehensible to us at the beginning of this study, but which has no longer anything mysterious about it, since we have ascertained the analogies between proof by recurrence and ordinary induction.

No doubt mathematical recurrent reasoning and physical inductive reasoning are based on different foundations, but they move in parallel lines and in the same direction — namely, from the particular to the general.

Let us examine the case a little more closely. To prove the equality $a + 2 = 2 + a$......(1), we need only apply the rule $a + 1 = 1 + a$, twice, and write $a + 2 = a + 1 + 1 = 1 + a + 1 = 1 + 1 + a = 2 + a$......(2).

The equality thus deduced by purely analytical means is not, however, a simple particular case. It is something quite different. We may not therefore even say in the really analytical and deductive part of mathematical reasoning that we proceed from the general to the particular in the ordinary sense of the words. The two sides of the equality (2) are merely more complicated combinations than the two sides of the equality (1), and analysis only serves to separate the elements which enter into these combinations and to study their relations.

Mathematicians therefore proceed "by construction," they "construct" more complicated combinations. When they analyse these combinations, these aggregates, so to speak, into their primitive elements, they see the relations of the elements and deduce the relations of the aggregates themselves. The process is purely analytical, but it is not a passing from the general to the particular, for the aggregates obviously cannot be regarded as more particular than their elements.

Great importance has been rightly attached to this process of "construction," and some claim to see in it the necessary and sufficient condition of the progress of the exact sciences. Necessary, no doubt, but not sufficient! For a construction to be useful and not mere waste of mental effort, for it to serve as a stepping-stone to higher things, it must first of all possess a kind of unity enabling us to see something more than the juxtaposition of its elements. Or more accurately, there must be some advantage in considering the construction rather than the elements themselves. What can this advantage be? Why reason on a polygon, for instance, which is always decomposable into triangles, and not on elementary triangles? It is because there are properties of polygons of any number of sides, and they can be immediately applied to any particular kind of polygon. In most cases it is only after long efforts that those properties can be discovered, by directly studying the relations of elementary triangles. If the quadrilateral is anything more than the juxtaposition of two triangles, it is because it is of the polygon type.

A construction only becomes interesting when it can be placed side by side with other analogous constructions for forming species of the same genus. To do this we must necessarily go back from the particular to the general, ascending one or more steps. The analytical process "by construction" does not compel us to descend, but it leaves us at the same level. We can only ascend by mathematical induction, for from it alone can we learn something new. Without the aid of this induction, which in certain respects differs from, but is as fruitful as, physical induction, construction would be powerless to create science.

Let me observe, in conclusion, that this induction is only possible if the same operation can be repeated indefinitely. That is why the theory of chess can never become a science, for the different moves of the same piece are limited and do not resemble each other.

Footnotes

1. For $(\gamma + 1) + 1 = (1 + \gamma) + 1 = 1 + (\gamma + 1)$. — [TR.]

CHAPTER II.
MATHEMATICAL MAGNITUDE AND EXPERIMENT.

IF we want to know what the mathematicians mean by a continuum, it is useless to appeal to geometry. The geometer is always seeking, more or less, to represent to himself the figures he is studying, but his representations are only instruments to him; he uses space in his geometry just as he uses chalk; and further, too much importance must not be attached to accidents which are often nothing more than the whiteness of the chalk.

The pure analyst has not to dread this pitfall. He has disengaged mathematics from all extraneous elements, and he is in a position to answer our question: — "Tell me exactly what this continuum is, about which mathematicians reason." Many analysts who reflect on their art have already done so — M. Tannery, for instance, in his *Introduction à la théorie des Fonctions d'une variable.*

Let us start with the integers. Between any two consecutive sets, intercalate one or more intermediary sets, and then between these sets others again, and so on indefinitely. We thus get an unlimited number of terms, and these will be the numbers which we call fractional, rational, or commensurable. But this is not yet all; between these terms, which, be it marked, are already infinite in number, other terms are intercalated, and these are called irrational or incommensurable.

Before going any further, let me make a preliminary remark. The continuum thus conceived is no longer a collection of individuals arranged in a certain order, infinite in number, it is true, but external the one to the other. This is not the ordinary conception in which it is supposed that between the elements of the continuum exists an intimate connection making of it one whole, in which the point has no existence previous to the line, but the line does exist previous to the point. Multiplicity alone subsists, unity has disappeared — "the continuum is unity in multiplicity," according to the celebrated formula. The analysts have even less reason to define their continuum as they do, since it is always on this that they reason when they are particularly proud of their rigour. It is enough to warn the reader that the real mathematical continuum is quite different from that of the physicists and from that of the metaphysicians.

It may also be said, perhaps, that mathematicians who are contented with this definition are the dupes of words, that the nature of each of these sets should be precisely indicated, that it should be explained how they are to be intercalated, and that it should be shown how it is possible to do it. This, however, would be wrong; the only property of the sets which comes into the reasoning is that of preceding or succeeding these or those other sets; this alone should therefore intervene in the definition. So we need not concern ourselves with the manner in which the sets are intercalated, and no one will doubt the possibility of the operation if he only remembers that "possible" in the language of geometers simply means exempt from contradiction. But our definition is not yet complete, and we come back to it after this rather long digression.

Definition of Incommensurables. — The mathematicians of the Berlin school, and Kronecker in particular, have devoted themselves to constructing this continuous scale of irrational and fractional numbers without using any other materials than the integer. The mathematical continuum from this point of view would be a pure creation of the mind in which experiment would have no part.

The idea of rational number not seeming to present to them any difficulty, they have confined their attention mainly to defining incommensurable numbers. But before reproducing their definition here, I must make an observation that will allay the astonishment which this will not fail to provoke in readers who are but little familiar with the habits of geometers.

Mathematicians do not study objects, but the relations between objects; to them it is a matter of indifference if these objects are replaced by others, provided that the relations do not change. Matter does not engage their attention, they are interested by form alone.

If we did not remember it, we could hardly understand that Kronecker gives the name of incommensurable number to a simple symbol — that is to say, something very different from the idea we think we ought to have of a quantity which should be measurable and almost tangible.

Let us see now what is Kronecker's definition. Commensurable numbers may be divided into classes in an infinite number of ways, subject to the condition that any number whatever of the first class is greater than any number of the second. It may happen that among the numbers of the first class there is one which is smaller than all the rest; if, for instance, we arrange in the first class all the numbers greater than 2, and 2 itself, and in the second class all the numbers smaller than 2, it is clear that 2 will be the smallest of all the numbers of the first class. The number 2 may therefore be chosen as the symbol of this division.

It may happen, on the contrary, that in the second class there is one which is greater than all the rest. This is what takes place, for example, if the first class comprises all the numbers greater than 2, and if, in the second, are all the numbers less than 2, and 2 itself. Here again the number 2 might be chosen as the symbol of this division.

But it may equally well happen that we can find neither in the first class a number smaller than all the rest, nor in the second class a number greater than all the rest. Suppose, for instance, we place in the first class all the numbers whose squares are greater than 2, and in the second all the numbers whose squares are smaller than 2. We know that in neither of them is a number whose square is equal to 2. Evidently there will be in the first class no number which is smaller than all the rest, for however near the square of a number may be to 2, we can always find a commensurable whose square is still nearer to 2. From Kronecker's point of view, the incommensurable number $\sqrt{2}$ is nothing but the symbol of this particular method of division of commensurable numbers; and to each mode of repartition corresponds in this way a number, commensurable or not, which serves as a symbol. But to be satisfied with this would be to forget the origin of these symbols; it remains to explain how we have been led to attribute to them a kind of concrete existence, and on the other hand, does not the difficulty begin with fractions? Should we have the notion of these numbers if we did not previously know a matter which we conceive as infinitely divisible — *i.e.*, as a continuum?

The Physical Continuum. — We are next led to ask if the idea of the mathematical continuum is not simply drawn from experiment. If that be so, the rough data of experiment, which are our sensations, could be measured. We might, indeed, be tempted to believe that this is so, for in recent times there has been an attempt to measure them, and a law has even been formulated, known as Fechner's law, according to which sensation is proportional to the logarithm of the stimulus. But if we examine the experiments by which the endeavour has been made to establish this law, we shall be led to a diametrically opposite conclusion. It has, for instance, been observed that a weight A of 10 grammes and a weight B of 11 grammes produced identical sensations, that the weight B could no longer be distinguished from a weight C of 12 grammes, but that the weight A was readily distinguished from the weight C. Thus the rough results of the experiments may be expressed by the following relations:

A = B, B = C, A < C, which may be regarded as the formula of the physical continuum. But here is an intolerable disagreement with the law of contradiction, and the necessity of banishing this disagreement has compelled us to invent the mathematical continuum. We are therefore forced to conclude that this notion has been created entirely by the mind, but it is experiment that has provided the opportunity. We cannot believe that two quantities which are equal to a third are not equal to one another, and we are thus led to suppose that A is different from B, and B from C, and that if we have not been aware of this, it is due to the imperfections of our senses.

The Creation of the Mathematical Continuum: First Stage. — So far it would suffice, in order to account for facts, to intercalate between A and B a small number of terms which would remain discrete. What happens now if we have recourse to some instrument to make up for the weakness of our senses? If, for example, we use a microscope? Such terms as A and B, which before were indistinguishable from one another, appear now to be distinct: but between A and B, which are distinct, is intercalated another new term D, which we can distinguish neither from A nor from B. Although we may use the most delicate methods, the rough results of our experiments will always present the characters of the physical continuum with the contradiction which is inherent in it. We only escape from it by incessantly intercalating new terms between the terms already distinguished, and this operation must be pursued indefinitely. We might conceive that it would be possible to stop if we could imagine an instrument powerful enough to decompose the physical continuum into discrete elements, just as the telescope resolves the Milky Way into stars. But this we cannot imagine; it is always with our senses that we use our instruments; it is with the eye that we observe the image magnified by the microscope, and this image must therefore always retain the characters of visual sensation, and therefore those of the physical continuum.

Nothing distinguishes a length directly observed from half that length doubled by the microscope. The whole is homogeneous to the part; and there is a fresh contradiction — or rather there would be one if the number of the terms were supposed to be finite; it is clear that the part containing less terms than the whole cannot be similar to the whole. The contradiction ceases as soon as the number of terms is regarded as infinite. There is nothing, for example, to prevent us from regarding the aggregate of integers as similar to the aggregate of even numbers, which is however only a part of it; in fact, to each integer corresponds another even number which is its double. But it is not only to escape this contradiction contained in the empiric data that the mind is led to create the concept of a continuum formed of an indefinite number of terms.

Here everything takes place just as in the series of the integers. We have the faculty of conceiving that a unit may be added to a collection of units. Thanks to experiment, we have had the opportunity of exercising this faculty and are conscious of it; but from this fact we feel that our power is unlimited, and that we can count indefinitely, although we have never had to count more than a finite number of objects. In the same way, as soon as we have intercalated terms between two consecutive terms of a series, we feel that this operation may be continued without limit, and that, so to speak, there is no intrinsic reason for stopping. As an abbreviation, I may give the name of a mathematical continuum of the first order to every aggregate of terms formed after the same law as the scale of commensurable numbers. If, then, we intercalate new sets according to the laws of incommensurable numbers, we obtain what may be called a continuum of the second order.

Second Stage. — We have only taken our first step. We have explained the origin of continuums of the first order; we must now see why this is not sufficient, and why the incommensurable numbers had to be invented.

If we try to imagine a line, it must have the characters of the physical continuum — that is to say, our representation must have a certain breadth. Two lines will therefore appear to us under the form of two narrow bands, and if we are content with this rough image, it is clear that where two lines cross they must have some common part. But the pure geometer makes one further effort; without entirely renouncing the aid of his senses, he tries to imagine a line without breadth and a point without size. This he can do only by imagining a line as the limit towards which tends a band that is growing thinner and thinner, and the point as the limit towards which is tending an area that is growing smaller and smaller. Our two bands, however narrow they may be, will always have a common area; the smaller they are the smaller it will be, and its limit is what the geometer calls a point. This is why it is said that the two lines which cross must have a common point, and this truth seems intuitive.

But a contradiction would be implied if we conceived of lines as continuums of the first order — *i.e.*, the lines traced by the geometer should only give us points, the co-ordinates of which are rational numbers. The contradiction would be manifest if we were, for instance, to assert the existence of lines and circles. It is clear, in fact, that if the points whose co-ordinates are commensurable were alone regarded as real, the in-circle of a square and the diagonal of the square would not intersect, since the co-ordinates of the point of intersection are incommensurable.

Even then we should have only certain incommensurable numbers, and not all these numbers.

But let us imagine a line divided into two half-rays (*demi-droites*). Each of these half-rays will appear to our minds as a band of a certain breadth; these bands will fit close together, because there must be no interval between them. The common part will appear to us to be a point which will still remain as we imagine the bands to become thinner and thinner, so that we admit as an intuitive truth that if a line be divided into two half-rays the common frontier of these half-rays is a point. Here we recognise the conception of Kronecker, in which an incommensurable number was regarded as the common frontier of two classes of rational numbers. Such is the origin of the continuum of the second order, which is the mathematical continuum properly so called.

Summary. — To sum up, the mind has the faculty of creating symbols, and it is thus that it has constructed the mathematical continuum, which is only a particular system of symbols. The only limit to its power is the necessity of avoiding all contradiction; but the mind only makes use of it when experiment gives a reason for it.

In the case with which we are concerned, the reason is given by the idea of the physical continuum, drawn from the rough data of the senses. But this idea leads to a series of contradictions from each of which in turn we must be freed. In this way we are forced to imagine a more and more complicated system of symbols. That on which we shall dwell is not merely exempt from internal contradiction, — it was so already at all the steps we have taken, — but it is no longer in contradiction with the various propositions which are called intuitive, and which are derived from more or less elaborate empirical notions.

Measurable Magnitude. — So far we have not spoken of the *measure* of magnitudes; we can tell if any one of them is greater than any other, but we cannot say that it is two or three times as large.

So far, I have only considered the order in which the terms are arranged; but that is not sufficient for most applications. We must learn how to compare the interval which separates any two terms. On this condition alone will the continuum become measurable, and the operations of arithmetic be applicable. This can only be done by the aid of a new and special convention; and this convention is, that in such a case the interval between the terms A and B is equal to the interval which separates C and D. For instance, we started with the integers, and between two consecutive sets we intercalated *n* intermediary sets; by convention we now assume these new sets to be equidistant. This is one of the ways of defining the addition of two magnitudes; for if the interval AB is by definition equal to the interval CD, the interval AD will by definition be the sum of the intervals AB and AC. This definition is very largely, but not altogether, arbitrary. It must satisfy certain conditions — the commutative and associative laws of addition, for instance; but, provided the definition we choose satisfies these laws, the choice is indifferent, and we need not state it precisely.

Remarks. — We are now in a position to discuss several important questions.

(1) Is the creative power of the mind exhausted by the creation of the mathematical continuum? The answer is in the negative, and this is shown in a very striking manner by the work of Du Bois Reymond.

We know that mathematicians distinguish between infinitesimals of different orders, and that infinitesimals of the second order are infinitely small, not only absolutely so, but also in relation to those of the first order. It is not difficult to imagine infinitesimals of fractional or even of irrational order, and here once more we find the mathematical continuum which has been dealt with in the preceding pages. Further, there are infinitesimals which are infinitely small with reference to those of the first order, and infinitely large with respect to the order $1 + \varepsilon$, however small ε may be. Here, then, are new terms intercalated in our series; and if I may be permitted to revert to the terminology used in the preceding pages, a terminology which is very convenient, although it has not been consecrated by usage, I shall say that we have created a kind of continuum of the third order.

It is an easy matter to go further, but it is idle to do so, for we would only be imagining symbols without any possible application, and no one will dream of doing that. This continuum of the third order, to which we are led by the consideration of the different orders of infinitesimals, is in itself of but little use and hardly worth quoting. Geometers look on it as a mere curiosity. The mind only uses its creative faculty when experiment requires it.

(2) When we are once in possession of the conception of the mathematical continuum, are we protected from contradictions analogous to those which gave it birth? No, and the following is an instance: —

He is a *savant* indeed who will not take it as evident that every curve has a tangent; and, in fact, if we think of a curve and a straight line as two narrow bands, we can always arrange them in such a way that they have a common part without intersecting. Suppose now that the breadth of the bands diminishes indefinitely: the common part will still remain, and in the limit, so to speak, the two lines will have a common point, although they do not intersect — *i.e.*, they will touch. The geometer who reasons in this way is only doing what we have done when we proved that two lines which intersect have a common point, and his intuition might also seem to be quite legitimate. But this is not the case. We can show that there are curves which have no tangent, if we define such a curve as an analytical continuum of the second order. No doubt some artifice analogous to those we have discussed above would enable us to get rid of this contradiction, but as the latter is only met with in very exceptional cases, we need not trouble to do so. Instead of endeavouring to reconcile intuition and analysis, we are content to sacrifice one of them, and as analysis must be flawless, intuition must go to the wall.

The Physical Continuum of several Dimensions. — We have discussed above the physical continuum as it is derived from the immediate evidence of our senses — or, if the reader prefers, from the rough results of Fechner's experiments; I have shown that these results are summed up in the contradictory formulae: — A = B, B = C, A < C.

Let us now see how this notion is generalised, and how from it may be derived the concept of continuums of several dimensions. Consider any two aggregates of sensations. We can either distinguish between them, or we cannot; just as in Fechner's experiments the weight of 10 grammes could be distinguished from the weight of 12 grammes, but not from the weight of 11 grammes. This is all that is required to construct the continuum of several dimensions.

Let us call one of these aggregates of sensations an *element*. It will be in a measure analogous to the *point* of the mathematicians, but will not be, however, the same thing. We cannot say that our element has no size, for we cannot distinguish it from its immediate neighbours, and it is thus surrounded by a kind of fog. If the astronomical comparison may be allowed, our "elements" would be like nebulae, whereas the mathematical points would be like stars.

If this be granted, a system of elements will form a continuum, if we can pass from any one of them to any other by a series of consecutive elements such that each cannot be distinguished from its predecessor. This *linear* series is to the *line* of the mathematician what the isolated *element* was to the point.

Before going further, I must explain what is meant by a *cut*. Let us consider a continuum C, and remove from it certain of its elements, which for a moment we shall regard as no longer belonging to the continuum. We shall call the aggregate of elements thus removed a *cut*. By means of this cut, the continuum C will be *subdivided* into several distinct continuums; the aggregate of elements which remain will cease to form a single continuum. There will then be on C two elements, A and B, which we must look upon as belonging to two distinct continuums; and we see that this must be so, because it will be impossible to find a linear series of consecutive elements of C (each of the elements indistinguishable from the preceding, the first being A and the last B), *unless one of the elements of this series is indistinguishable from one of the elements of the cut.*

It may happen, on the contrary, that the cut may not be sufficient to subdivide the continuum C. To classify the physical continuums, we must first of all ascertain the nature of the cuts which must be made in order to subdivide them. If a physical continuum, C, may be subdivided by a cut reducing to a finite number of elements, all distinguishable the one from the other (and therefore forming neither one continuum nor several continuums), we shall call C a continuum *of one dimension*. If, on the contrary, C can only be subdivided by cuts which are themselves continuums, we shall say that C is of several dimensions; if the cuts are continuums of one dimension, then we shall say that C has two dimensions; if cuts of two dimensions are sufficient, we shall say that C is of three dimensions, and so on. Thus the notion of the physical continuum of several dimensions is defined, thanks to the very simple fact, that two aggregates of sensations may be distinguishable or indistinguishable.

The Mathematical Continuum of Several Dimensions. — The conception of the mathematical continuum of n dimensions may be led up to quite naturally by a process similar to that which we discussed at the beginning of this chapter. A point of such a continuum is defined by a system of n distinct magnitudes which we call its co-ordinates.

The magnitudes need not always be measurable; there is, for instance, one branch of geometry independent of the measure of magnitudes, in which we are only concerned with knowing, for example, if, on a curve ABC, the point B is between the points A and C, and in which it is immaterial whether the arc AB is equal to or twice the arc BC. This branch is called *Analysis Situs*. It contains quite a large body of doctrine which has attracted the attention of the greatest geometers, and from which are derived, one from another, a whole series of remarkable theorems. What distinguishes these theorems from those of ordinary geometry is that they are purely qualitative. They are still true if the figures are copied by an unskilful draughtsman, with the result that the proportions are distorted and the straight lines replaced by lines which are more or less curved.

As soon as measurement is introduced into the continuum we have just defined, the continuum becomes space, and geometry is born. But the discussion of this is reserved for Part II.

PART II.

SPACE.

CHAPTER III.
NON-EUCLIDEAN GEOMETRIES.

EVERY conclusion presumes premisses. These premisses are either self-evident and need no demonstration, or can be established only if based on other propositions; and, as we cannot go back in this way to infinity, every deductive science, and geometry in particular, must rest upon a certain number of indemonstrable axioms. All treatises of geometry begin therefore with the enunciation of these axioms. But there is a distinction to be drawn between them. Some of these, for example, "Things which are equal to the same thing are equal to one another," are not propositions in geometry but propositions in analysis. I look upon them as analytical à priori intuitions, and they concern me no further. But I must insist on other axioms which are special to geometry. Of these most treatises explicitly enunciate three: — (1) Only one line can pass through two points; (2) a straight line is the shortest distance between two points; (3) through one point only one parallel can be drawn to a given straight line. Although we generally dispense with proving the second of these axioms, it would be possible to deduce it from the other two, and from those much more numerous axioms which are implicitly admitted without enunciation, as I shall explain further on. For a long time a proof of the third axiom known as Euclid's postulate was sought in vain. It is impossible to imagine the efforts that have been spent in pursuit of this chimera. Finally, at the beginning of the nineteenth century, and almost simultaneously, two scientists, a Russian and a Bulgarian, Lobatschewsky and Bolyai, showed irrefutably that this proof is impossible. They have nearly rid us of inventors of geometries without a postulate, and ever since the Académie des Sciences receives only about one or two new demonstrations a year. But the question was not exhausted, and it was not long before a great step was taken by the celebrated memoir of Riemann, entitled: *Ueber die Hypothesen welche der Geometrie zum Grunde liegen.* This little work has inspired most of the recent treatises to which I shall later on refer, and among which I may mention those of Beltrami and Helmholtz.

The Geometry of Lobatschewsky. — If it were possible to deduce Euclid's postulate from the several axioms, it is evident that by rejecting the postulate and retaining the other axioms we should be led to contradictory consequences. It would be, therefore, impossible to found on those premisses a coherent geometry. Now, this is precisely what Lobatschewsky has done. He assumes at the outset that several parallels may be drawn through a point to a given straight line, and he retains all the other axioms of Euclid. From these hypotheses he deduces a series of theorems between which it is impossible to find any contradiction, and he constructs a geometry as impeccable in its logic as Euclidean geometry. The theorems are very different, however, from those to which we are accustomed, and at first will be found a little disconcerting. For instance, the sum of the angles of a triangle is always less than two right angles, and the difference between that sum and two right angles is proportional to the area of the triangle. It is impossible to construct a figure similar to a given figure but of different dimensions. If the circumference of a circle be divided into n equal parts, and tangents be drawn at the points of intersection, the n tangents will form a polygon if the radius of the circle is small enough, but if the radius is large enough they will never meet. We need not multiply these examples. Lobatschewsky's propositions have no relation to those of Euclid, but they are none the less logically interconnected.

Riemann's Geometry. — Let us imagine to ourselves a world only peopled with beings of no thickness, and suppose these "infinitely flat" animals are all in one and the same plane, from which they cannot emerge. Let us further admit that this world is sufficiently distant from other worlds to be withdrawn from their influence, and while we are making these hypotheses it will not cost us much to endow these beings with reasoning power, and to believe them capable of making a geometry. In that case they will certainly attribute to space only two dimensions. But now suppose that these imaginary animals, while remaining without thickness, have the form of a spherical, and not of a plane figure, and are all on the same sphere, from which they cannot escape. What kind of a geometry will they construct? In the first place, it is clear that they will attribute to space only two dimensions. The straight line to them will be the shortest distance from one point on the sphere to another — that is to say, an arc of a great circle. In a word, their geometry will be spherical geometry. What they will call space will be the sphere on which they are confined, and on which take place all the phenomena with which they are acquainted. Their space will therefore be *unbounded*, since on a sphere one may always walk forward without ever being brought to a stop, and yet it will be *finite*; the end will never be found, but the complete tour can be made. Well, Riemann's geometry is spherical geometry extended to three dimensions. To construct it, the German mathematician had first of all to throw overboard, not only Euclid's postulate but also the first axiom that *only one line can pass through two points*. On a sphere, through two given points, we can *in general* draw only one great circle which, as we have just seen, would be to our imaginary beings a straight line. But there was one exception. If the two given points are at the ends of a diameter, an infinite number of great circles can be drawn through them. In the same way, in Riemann's geometry — at least in one of its forms — through two points only one straight line can in general be drawn, but there are exceptional cases in which through two points an infinite number of straight lines can be drawn. So there is a kind of opposition between the geometries of Riemann and Lobatschewsky. For instance, the sum of the angles of a triangle is equal to two right angles in Euclid's geometry, less than two right angles in that of Lobatschewsky, and greater than two right angles in that of Riemann. The number of parallel lines that can be drawn through a given point to a given line is one in Euclid's geometry, none in Riemann's, and an infinite number in the geometry of Lobatschewsky. Let us add that Riemann's space is finite, although unbounded in the sense which we have above attached to these words.

Surfaces with Constant Curvature. — One objection, however, remains possible. There is no contradiction between the theorems of Lobatschewsky and Riemann; but however numerous are the other consequences that these geometers have deduced from their hypotheses, they had to arrest their course before they exhausted them all, for the number would be infinite; and who can say that if they had carried their deductions further they would not have eventually reached some contradiction? This difficulty does not exist for Riemann's geometry, provided it is limited to two dimensions. As we have seen, the two-dimensional geometry of Riemann, in fact, does not differ from spherical geometry, which is only a branch of ordinary geometry, and is therefore outside all contradiction. Beltrami, by showing that Lobatschewsky's two-dimensional geometry was only a branch of ordinary geometry, has equally refuted the objection as far as it is concerned. This is the course of his argument: Let us consider any figure whatever on a surface. Imagine this figure to be traced on a flexible and inextensible canvas applied to the surface, in such a way that when the canvas is displaced and deformed the different lines of the figure change their form without changing their length. As a rule, this flexible and inextensible figure cannot be displaced without leaving the surface. But there are certain surfaces for which such a movement would be possible. They are surfaces of constant curvature. If we resume the comparison that we made just now, and imagine beings without thickness living on one of these surfaces, they will regard as possible the motion of a figure all the lines of which remain of a constant length. Such a movement would appear absurd, on the other hand, to animals without thickness living on a surface of variable curvature. These surfaces of constant curvature are of two kinds. The curvature of some is *positive*, and they may be deformed so as to be applied to a sphere. The geometry of these surfaces is therefore reduced to spherical geometry — namely, Riemann's. The curvature of others is *negative*. Beltrami has shown that the geometry of these surfaces is identical with that of Lobatschewsky. Thus the two-dimensional geometries of Riemann and Lobatschewsky are connected with Euclidean geometry.

Interpretation of Non-Euclidean Geometries. — Thus vanishes the objection so far as two-dimensional geometries are concerned. It would be easy to extend Beltrami's reasoning to three-dimensional geometries, and minds which do not recoil before space of four dimensions will see no difficulty in it; but such minds are few in number. I prefer, then, to proceed otherwise. Let us consider a certain plane, which I shall call the fundamental plane, and let us construct a kind of dictionary by making a double series of terms written in two columns, and corresponding each to each, just as in ordinary dictionaries the words in two languages which have the same signification correspond to one another: —

Space	The portion of space situated above the fundamental plane.
Plane	Sphere cutting orthogonally the fundamental plane.
Line	Circle cutting orthogonally the fundamental plane.
Sphere	Sphere.
Circle	Circle.
Angle	Angle.
Distance between two points ...	Logarithm of the anharmonic ratio of these two points and of the intersection of the fundamental plane with the circle passing through these two points and cutting it orthogonally.
Etc.	Etc.

Let us now take Lobatschewsky's theorems and translate them by the aid of this dictionary, as we would translate a German text with the aid of a German - French dictionary. *We shall then obtain the theorems of ordinary geometry.* For instance, Lobatschewsky's theorem: "The sum of the angles of a triangle is less than two right angles," may be translated thus: "If a curvilinear triangle has for its sides arcs of circles which if produced would cut orthogonally the fundamental plane, the sum of the angles of this curvilinear triangle will be less than two right angles." Thus, however far the consequences of Lobatschewsky's hypotheses are carried, they will never lead to a contradiction; in fact, if two of Lobatschewsky's theorems were contradictory, the translations of these two theorems made by the aid of our dictionary would be contradictory also. But these translations are theorems of ordinary geometry, and no one doubts that ordinary geometry is exempt from contradiction. Whence is the certainty derived, and how far is it justified? That is a question upon which I cannot enter here, but it is a very interesting question, and I think not insoluble. Nothing, therefore, is left of the objection I formulated above. But this is not all. Lobatschewsky's geometry being susceptible of a concrete interpretation, ceases to be a useless logical exercise, and may be applied. I have no time here to deal with these applications, nor with what Herr Klein and myself have done by using them in the integration of linear equations. Further, this interpretation is not unique, and several dictionaries may be constructed analogous to that above, which will enable us by a simple translation to convert Lobatschewsky's theorems into the theorems of ordinary geometry.

Implicit Axioms. — Are the axioms implicitly enunciated in our text-books the only foundation of geometry? We may be assured of the contrary when we see that, when they are abandoned one after another, there are still left standing some propositions which are common to the geometries of Euclid, Lobatschewsky, and Riemann. These propositions must be based on premisses that geometers admit without enunciation. It is interesting to try and extract them from the classical proofs.

John Stuart Mill asserted[1] that every definition contains an axiom, because by defining we implicitly affirm the existence of the object defined. That is going rather too far. It is but rarely in mathematics that a definition is given without following it up by the proof of the existence of the object defined, and when this is not done it is generally because the reader can easily supply it; and it must not be forgotten that the word "existence" has not the same meaning when it refers to a mathematical entity as when it refers to a material object.

A mathematical entity exists provided there is no contradiction implied in its definition, either in itself, or with the propositions previously admitted. But if the observation of John Stuart Mill cannot be applied to all definitions, it is none the less true for some of them. A plane is sometimes defined in the following manner: — The plane is a surface such that the line which joins any two points upon it lies wholly on that surface. Now, there is obviously a new axiom concealed in this definition. It is true we might change it, and that would be preferable, but then we should have to enunciate the axiom explicitly. Other definitions may give rise to no less important reflections, such as, for example, that of the equality of two figures. Two figures are equal when they can be superposed. To superpose them, one of them must be displaced until it coincides with the other. But how must it be displaced? If we asked that question, no doubt we should be told that it ought to be done without deforming it, and as an invariable solid is displaced. The vicious circle would then be evident. As a matter of fact, this definition defines nothing. It has no meaning to a being living in a world in which there are only fluids. If it seems clear to us, it is because we are accustomed to the properties of natural solids which do not much differ from those of the ideal solids, all of whose dimensions are invariable. However, imperfect as it may be, this definition implies an axiom. The possibility of the motion of an invariable figure is not a self-evident truth. At least it is only so in the application to Euclid's postulate, and not as an analytical à priori intuition would be. Moreover, when we study the definitions and the proofs of geometry, we see that we are compelled to admit without proof not only the possibility of this motion, but also some of its properties. This first arises in the definition of the straight line. Many defective definitions have been given, but the true one is that which is understood in all the proofs in which the straight line intervenes. "It may happen that the motion of an invariable figure may be such that all the points of a line belonging to the figure are motionless, while all the points situate outside that line are in motion. Such a line would be called a straight line." We have deliberately in this enunciation separated the definition from the axiom which it implies. Many proofs such as those of the cases of the equality of triangles, of the possibility of drawing a perpendicular from a point to a straight line, assume propositions the enunciations of which are dispensed with, for they necessarily imply that it is possible to move a figure in space in a certain way.

The Fourth Geometry. — Among these explicit axioms there is one which seems to me to deserve some attention, because when we abandon it we can construct a fourth geometry as coherent as those of Euclid, Lobatschewsky, and Riemann. To prove that we can always draw a perpendicular at a point A to a straight line A B, we consider a straight line A C movable about the point A, and initially identical with the fixed straight line A B. We then can make it turn about the point A until it lies in A B produced. Thus we assume two propositions — first, that such a rotation is possible, and then that it may continue until the two lines lie the one in the other produced. If the first point is conceded and the second rejected, we are led to a series of theorems even stranger than those of Lobatschewsky and Riemann, but equally free from contradiction. I shall give only one of these theorems, and I shall not choose the least remarkable of them. *A real straight line may be perpendicular to itself.*

Lie's Theorem. — The number of axioms implicitly introduced into classical proofs is greater than necessary, and it would be interesting to reduce them to a minimum. It may be asked, in the first place, if this reduction is possible — if the number of necessary axioms and that of imaginable geometries is not infinite? A theorem due to Sophus Lie is of weighty importance in this discussion. It may be enunciated in the following manner: — Suppose the following premises are admitted: (1) space has *n* dimensions; (2) the movement of an invariable figure is possible; (3) *p* conditions are necessary to determine the position of this figure in space.

The number of geometries compatible with these premises will be limited. I may even add that if *n* is given, a superior limit can be assigned to *p*. If, therefore, the possibility of the movement is granted, we can only invent a finite and even a rather restricted number of three-dimensional geometries.

Riemann's Geometries. — However, this result seems contradicted by Riemann, for that scientist constructs an infinite number of geometries, and that to which his name is usually attached is only a particular case of them. All depends, he says, on the manner in which the length of a curve is defined. Now, there is an infinite number of ways of defining this length, and each of them may be the starting-point of a new geometry. That is perfectly true, but most of these definitions are incompatible with the movement of a variable figure such as we assume to be possible in Lie's theorem. These geometries of Riemann, so interesting on various grounds, can never be, therefore, purely analytical, and would not lend themselves to proofs analogous to those of Euclid.

On the Nature of Axioms. — Most mathematicians regard Lobatschewsky's geometry as a mere logical curiosity. Some of them have, however, gone further. If several geometries are possible, they say, is it certain that our geometry is the one that is true? Experiment no doubt teaches us that the sum of the angles of a triangle is equal to two right angles, but this is because the triangles we deal with are too small. According to Lobatschewsky, the difference is proportional to the area of the triangle, and will not this become sensible when we operate on much larger triangles, and when our measurements become more accurate? Euclid's geometry would thus be a provisory geometry. Now, to discuss this view we must first of all ask ourselves, what is the nature of geometrical axioms? Are they synthetic *à priori* intuitions, as Kant affirmed? They would then be imposed upon us with such a force that we could not conceive of the contrary proposition, nor could we build upon it a theoretical edifice. There would be no non-Euclidean geometry. To convince ourselves of this, let us take a true synthetic *à priori* intuition — the following, for instance, which played an important part in the first chapter: — If a theorem is true for the number 1, and if it has been proved that it is true of *n* + 1, provided it is true of *n*, it will be true for all positive integers. Let us next try to get rid of this, and while rejecting this proposition let us construct a false arithmetic analogous to non-Euclidean geometry. We shall not be able to do it. We shall be even tempted at the outset to look upon these intuitions as analytical. Besides, to take up again our fiction of animals without thickness, we can scarcely admit that these beings, if their minds are like ours, would adopt the Euclidean geometry, which would be contradicted by all their experience. Ought we, then, to conclude that the axioms of geometry are experimental truths? But we do not make experiments on ideal lines or ideal circles; we can only make them on material objects. On what, therefore, would experiments serving as a foundation for geometry be based? The answer is easy. We have seen above that we constantly reason as if the geometrical figures behaved like solids. What geometry would borrow from experiment would be therefore the properties of these bodies. The properties of light and its propagation in a straight line have also given rise to some of the propositions of geometry, and in particular to those of projective geometry, so that from that point of view one would be tempted to say that metrical geometry is the study of solids, and projective geometry that of light. But a difficulty remains, and is unsurmountable. If geometry were an experimental science, it would not be an exact science. It would be subjected to continual revision. Nay, it would from that day forth be proved to be erroneous, for we know that no rigorously invariable solid exists. *The geometrical axioms are therefore neither synthetic à priori intuitions nor experimental facts.* They are conventions. Our choice among all possible conventions is *guided* by experimental facts; but it remains *free*, and is only limited by the necessity of avoiding every contradiction, and thus it is that postulates may remain rigorously true even when the experimental laws which have determined their adoption are only approximate. In other words, *the axioms of geometry* (I do not speak of those of arithmetic) *are only definitions in disguise.* What, then, are we to think of the question: Is Euclidean geometry true? It has no meaning. We might as well ask if the metric system is true, and if the old weights and measures are false; if Cartesian co-ordinates are true and polar co-ordinates false. One geometry cannot be more true than another; it can only be more convenient. Now, Euclidean geometry is, and will remain, the most convenient: 1st, because it is the simplest, and it is not so only because of our mental habits or because of the kind of direct intuition that we have of Euclidean space; it is the simplest in itself, just as a polynomial of the first degree is simpler than a polynomial of the second degree; 2nd, because it sufficiently agrees with the properties of natural solids, those bodies which we can compare and measure by means of our senses.

Footnotes

1. *Logic*, c. viii., cf. Definitions, § 5-6. Tr.

CHAPTER IV.
SPACE AND GEOMETRY.

LET us begin with a little paradox. Beings whose minds were made as ours, and with senses like ours, but without any preliminary education, might receive from a suitably-chosen external world impressions which would lead them to construct a geometry other than that of Euclid, and to localise the phenomena of this external world in a non-Euclidean space, or even in space of four dimensions. As for us, whose education has been made by our actual world, if we were suddenly transported into this new world, we should have no difficulty in referring phenomena to our Euclidean space. Perhaps somebody may appear on the scene some day who will devote his life to it, and be able to represent to himself the fourth dimension.

Geometrical Space and Representative Space. — It is often said that the images we form of external objects are localised in space, and even that they can only be formed on this condition. It is also said that this space, which thus serves as a kind of framework ready prepared for our sensations and representations, is identical with the space of thegeometers, having all the properties of that space. To all clear-headed men who think in this way, the preceding statement might well appear extraordinary; but it is as well to see if they are not the victims of some illusion which closer analysis may be able to dissipate. In the first place, what are the properties of space properly so called? I mean of that space which is the object of geometry, and which I shall call geometrical space. The following are some of the more essential:—

1st, it is continuous; 2nd, it is infinite; 3rd, it is of three dimensions; 4th, it is homogeneous — that is to say, all its points are identical one with another; 5th, it is isotropic. Compare this now with the framework of our representations and sensations, which I may call *representative space*.

Visual Space. — First of all let us consider a purely visual impression, due to an image formed on the back of the retina. A cursory analysis shows us this image as continuous, but as possessing only two dimensions, which already distinguishes purely visual from what may be called geometrical space. On the other hand, the image is enclosed within a limited framework; and there is a no less important difference: *this pure visual space is not homogeneous*. All the points on the retina, apart from the images which may be formed, do not play the same role. The yellow spot can in no way be regarded as identical with a point on the edge of the retina. Not only does the same object produce on it much brighter impressions, but in the whole of the *limited* framework the point which occupies the centre will not appear identical with a point near one of the edges. Closer analysis no doubt would show us that this continuity of visual space and its two dimensions are but an illusion. It would make visual space even more different than before from geometrical space, but we may treat this remark as incidental.

However, sight enables us to appreciate distance, and therefore to perceive a third dimension. But every one knows that this perception of the third dimension reduces to a sense of the effort of accommodation which must be made, and to a sense of the convergence of the two eyes, that must take place in order to perceive an object distinctly. These are muscular sensations quite different from the visual sensations which have given us the concept of the two first dimensions. The third dimension will therefore not appear to us as playing the same rôle as the two others. What may be called *complete visual space* is not therefore an isotropic space. It has, it is true, exactly three dimensions; which means that the elements of our visual sensations (those at least which concur in forming the concept of extension) will be completely defined if we know three of them; or, in mathematical language, they will be functions of three independent variables. But let us look at the matter a little closer. The third dimension is revealed to us in two different ways: by the effort of accommodation, and by the convergence of the eyes. No doubt these two indications are always in harmony; there is between them a constant relation; or, in mathematical language, the two variables which measure these two muscular sensations do not appear to us as independent. Or, again, to avoid an appeal to mathematical ideas which are already rather too refined, we may go back to the language of the preceding chapter and enunciate the same fact as follows: — If two sensations of convergence A and B are indistinguishable, the two sensations of accommodation A' and B' which accompany them respectively will also be indistinguishable. But that is, so to speak, an experimental fact. Nothing prevents us *à priori* from assuming the contrary, and if the contrary takes place, if these two muscular sensations both vary independently, we must take into account one more independent variable, and complete visual space will appear to us as a physical continuum of four dimensions. And so in this there is also a fact of *external* experiment. Nothing prevents us from assuming that a being with a mind like ours, with the same sense-organs as ourselves, may be placed in a world in which light would only reach him after being passed through refracting media of complicated form. The two indications which enable us to appreciate distances would cease to be connected by a constant relation. A being educating his senses in such a world would no doubt attribute four dimensions to complete visual space.

Tactile and Motor Space. — "Tactile space" is more complicated still than visual space, and differs even more widely from geometrical space. It is useless to repeat for the sense of touch my remarks on the sense of sight. But outside the data of sight and touch there are other sensations which contribute as much and more than they do to the genesis of the concept of space. They are those which everybody knows, which accompany all our movements, and which we usually call muscular sensations. The corresponding framework constitutes what may be called *motor space*. Each muscle gives rise to a special sensation which may be increased or diminished so that the aggregate of our muscular sensations will depend upon as many variables as we have muscles. From this point of view *motor space would have as many dimensions as we have muscles*. I know that it is said that if the muscular sensations contribute to form the concept of space, it is because we have the sense of the *direction* of each movement, and that this is an integral part of the sensation. If this were so, and if a muscular sense could not be aroused unless it were accompanied by this geometrical sense of direction, geometrical space would certainly be a form imposed upon our sensitiveness. But I do not see this at all when I analyse my sensations. What I do see is that the sensations which correspond to movements in the same direction are connected in my mind by a simple *association of ideas*. It is to this association that what we call the sense of direction is reduced. We cannot therefore discover this sense in a single sensation. This association is extremely complex, for the contraction of the same muscle may correspond, according to the position of the limbs, to very different movements of direction. Moreover, it is evidently acquired; it is like all associations of ideas, the result of a *habit*. This habit itself is the result of a very large number of *experiments*, and no doubt if the education of our senses had taken place in a different medium, where we would have been subjected to different impressions, then contrary habits would have been acquired, and our muscular sensations would have been associated according to other laws.

Characteristics of Representative Space. — Thus representative space in its triple form — visual, tactile, and motor — differs essentially from geometrical space. It is neither homogeneous nor isotropic; we cannot even say that it is of three dimensions. It is often said that we "project" into geometrical space the objects of our external perception; that we "localise" them. Now, has that any meaning, and if so what is that meaning? Does it mean that we *represent* to ourselves external objects in geometrical space? Our representations are only the reproduction of our sensations; they cannot therefore be arranged in the same framework — that is to say, in representative space. It is also just as impossible for us to represent to ourselves external objects in geometrical space, as it is impossible for a painter to paint on a flat surface objects with their three dimensions. Representative space is only an image of geometrical space, an image deformed by a kind of perspective, and we can only represent to ourselves objects by making them obey the laws of this perspective. Thus we do not *represent* to ourselves external bodies in geometrical space, but we *reason* about these bodies as if they were situated in geometrical space. When it is said, on the other hand, that we "localise" such an object in such a point of space, what does it mean? *It simply means that we represent to ourselves the movements that must take place to reach that object.* And it does not mean that to represent to ourselves these movements they must be projected into space, and that the concept of space must therefore pre-exist. When I say that we represent to ourselves these movements, I only mean that we represent to ourselves the muscular sensations which accompany them, and which have no geometrical character, and which therefore in no way imply the pre-existence of the concept of space.

Changes of State and Changes of Position. — But, it may be said, if the concept of geometrical space is not imposed upon our minds, and if, on the other hand, none of our sensations can furnish us with that concept, how then did it ever come into existence? This is what we have now to examine, and it will take some time; but I can sum up in a few words the attempt at explanation which I am going to develop. *None of our sensations, if isolated, could have brought us to the concept of space; we are brought to it solely by studying the laws by which those sensations succeed one another.* We see at first that our impressions are subject to change; but among the changes that we ascertain, we are very soon led to make a distinction. Sometimes we say that the objects, the causes of these impressions, have changed their state, sometimes that they have changed their position, that they have only been displaced. Whether an object changes its state or only its position, this is always translated for us in the same manner, *by a modification in an aggregate of impressions.* How then have we been enabled to distinguish them? If there were only change of position, we could restore the primitive aggregate of impressions by making movements which would confront us with the movable object in the same *relative* situation. We thus *correct* the modification which was produced, and we re-establish the initial state by an inverse modification. If, for example, it were a question of the sight, and if an object be displaced before our eyes, we can "follow it with the eye," and retain its image on the same point of the retina by appropriate movements of the eyeball. These movements we are conscious of because they are voluntary, and because they are accompanied by muscular sensations. But that does not mean that we represent them to ourselves in geometrical space. So what characterises change of position, what distinguishes it from change of state, is that it can always be *corrected* by this means. It may therefore happen that we pass from the aggregate of impressions A to the aggregate B in two different ways. First, involuntarily and without experiencing muscular sensations — which happens when it is the object that is displaced; secondly, voluntarily, and with muscular sensation — which happens when the object is motionless, but when we displace ourselves in such a way that the object has relative motion with respect to us. If this be so, the translation of the aggregate A to the aggregate B is only a change of position. It follows that sight and touch could not have given us the idea of space without the help of the "muscular sense." Not only could this concept not be derived from a single sensation, or even from *a series of sensations*; but a *motionless* being could never have acquired it, because, not being able to correct by his movements the effects of the change of position of external objects, he would have had no reason to distinguish them from changes of state. Nor would he have been able to acquire it if his movements had not been voluntary, or if they were unaccompanied by any sensations whatever.

Conditions of Compensation. — How is such a compensation possible in such a way that two changes, otherwise mutually independent, may be reciprocally corrected? A mind *already familiar with geometry* would reason as follows: — If there is to be compensation, the different parts of the external object on the one hand, and the different organs of our senses on the other, must be in the same *relative* position after the double change. And for that to be the case, the different parts of the external body on the one hand, and the different organs of our senses on the other, must have the same relative position to each other after the double change; and so with the different parts of our body with respect to each other. In other words, the external object in the first change must be displaced as an invariable solid would be displaced, and it must also be so with the whole of our body in the second change, which is to correct the first. Under these conditions compensation may be produced. But we who as yet know nothing of geometry, whose ideas of space are not yet formed, we cannot reason in this way — we cannot predict *à priori* if compensation is possible. But experiment shows us that it sometimes does take place, and we start from this experimental fact in order to distinguish changes of state from changes of position.

Solid Bodies and Geometry. — Among surrounding objects there are some which frequently experience displacements that may be thus corrected by a *correlative* movement of our own body — namely, *solid bodies.* The other objects, whose form is variable, only in exceptional circumstances undergo similar displacement (change of position without change of form). When the displacement of a body takes place with deformation, we can no longer by appropriate movements place the organs of our body in the same *relative* situation with respect to this body; we can no longer, therefore, reconstruct the primitive aggregate of impressions.

It is only later, and after a series of new experiments, that we learn how to decompose a body of variable form into smaller elements such that each is displaced approximately according to the same laws as solid bodies. We thus distinguish "deformations" from other changes of state. In these deformations each element undergoes a simple change of position which may be corrected; but the modification of the aggregate is more profound, and can no longer be corrected by a correlative movement. Such a concept is very complex even at this stage, and has been relatively slow in its appearance. It would not have been conceived at all had not the observation of solid bodies shown us beforehand how to distinguish changes of position.

If, then, there were no solid bodies in nature there would be no geometry.

Another remark deserves a moment's attention. Suppose a solid body to occupy successively the positions α and β; in the first position it will give us an aggregate of impressions A, and in the second position the aggregate of impressions B. Now let there be a second solid body, of qualities entirely different from the first — of different colour, for instance. Assume it to pass from the position α, where it gives us the aggregate of impressions A' to the position β, where it gives the aggregate of impressions B'. In general, the aggregate A will have nothing in common with the aggregate A', nor will the aggregate B have anything in common with the aggregate B'. The transition from the aggregate A to the aggregate B, and that of the aggregate A' to the aggregate B', are therefore two changes which *in themselves* have in general nothing in common. Yet we consider both these changes as displacements; and, further, we consider them the *same* displacement. How can this be? It is simply because they may be both corrected by the *same* correlative movement of our body. "Correlative movement," therefore, constitutes the *sole connection* between two phenomena which otherwise we should never have dreamed of connecting.

On the other hand, our body, thanks to the number of its articulations and muscles, may have a multitude of different movements, but all are not capable of "correcting" a modification of external objects; those alone are capable of it in which our whole body, or at least all those in which the organs of our senses enter into play are displaced *en bloc*— *i.e.*, without any variation of their relative positions, as in the case of a solid body.

To sum up:

1. In the first place, we distinguish two categories of phenomena: — The first involuntary, unaccompanied by muscular sensations, and attributed to external objects — they are external changes; the second, of opposite character and attributed to the movements of our own body, are internal changes.

2. We notice that certain changes of each in these categories may be corrected by a correlative change of the other category.

3. We distinguish among external changes those that have a correlative in the other category — which we call displacements; and in the same way we distinguish among the internal changes those which have a correlative in the first category.

Thus by means of this reciprocity is defined a particular class of phenomena called displacements. *The laws of these phenomena are the object of geometry.*

Law of Homogeneity. — The first of these laws is the law of homogeneity. Suppose that by an external change we pass from the aggregate of impressions A to the aggregate B, and that then this change α is corrected by a correlative voluntary movement β, so that we are brought back to the aggregate A. Suppose now that another external change α' brings us again from the aggregate A to the aggregate B. Experiment then shows us that this change α', like the change α, may be corrected by a voluntary correlative movement β', and that this movement β' corresponds to the same muscular sensations as the movement β which corrected α.

This fact is usually enunciated as follows: — *Space is homogeneous and isotropic.* We may also say that a movement which is once produced may be repeated a second and a third time, and so on, without any variation of its properties. In the first chapter, in which we discussed the nature of mathematical reasoning, we saw the importance that should be attached to the possibility of repeating the same operation indefinitely. The virtue of mathematical reasoning is due to this repetition; by means of the law of homogeneity geometrical facts are apprehended. To be complete, to the law of homogeneity must be added a multitude of other laws, into the details of which I do not propose to enter, but which mathematicians sum up by saying that these displacements form a "group."

The Non-Euclidean World. — If geometrical space were a framework imposed on *each* of our representations considered individually, it would be impossible to represent to ourselves an image without this framework, and we should be quite unable to change our geometry. But this is not the case; geometry is only the summary of the laws by which these images succeed each other. There is nothing, therefore, to prevent us from imagining a series of representations, similar in every way to our ordinary representations, but succeeding one another according to laws which differ from those to which we are accustomed. We may thus conceive that beings whose education has taken place in a medium in which those laws would be so different, might have a very different geometry from ours.

Suppose, for example, a world enclosed in a large sphere and subject to the following laws: — The temperature is not uniform; it is greatest at the centre, and gradually decreases as we move towards the circumference of the sphere, where it is absolute zero. The law of this temperature is as follows: If

R be the radius of the sphere, and r the distance of the point considered from the centre, the absolute temperature will be proportional to R^2-r^2. Further, I shall suppose that in this world all bodies have the same co-efficient of dilatation, so that the linear dilatation of any body is proportional to its absolute temperature. Finally, I shall assume that a body transported from one point to another of different temperature is instantaneously in thermal equilibrium with its new environment. There is nothing in these hypotheses either contradictory or unimaginable. A moving object will become smaller and smaller as it approaches the circumference of the sphere. Let us observe, in the first place, that although from the point of view of our ordinary geometry this world is finite, to its inhabit ants it will appear infinite. As they approach the surface of the sphere they become colder, and at the same time smaller and smaller. The steps they take are therefore also smaller and smaller, so that they can never reach the boundary of the sphere. If to us geometry is only the study of the laws according to which invariable solids move, to these imaginary beings it will be the study of the laws of motion of solids *deformed by the differences of temperature* alluded to.

No doubt, in our world, natural solids also experience variations of form and volume due to differences of temperature. But in laying the foundations of geometry we neglect these variations; for besides being but small they are irregular, and consequently appear to us to be accidental. In our hypothetical world this will no longer be the case, the variations will obey very simple and regular laws. On the other hand, the different solid parts of which the bodies of these inhabitants are composed will undergo the same variations of form and volume.

Let me make another hypothesis: suppose that light passes through media of different refractive indices, such that the index of refraction is inversely proportional to R^2-r^2. Under these conditions it is clear that the rays of light will no longer be rectilinear but circular. To justify what has been said, we have to prove that certain changes in the position of external objects may be corrected by correlative movements of the beings which inhabit this imaginary world; and in such a way as to restore the primitive aggregate of the impressions experienced by these sentient beings. Suppose, for example, that an object is displaced and deformed, not like an invariable solid, but like a solid subjected to unequal dilatations in exact conformity with the law of temperature assumed above. To use an abbreviation, we shall call such a movement a non-Euclidean displacement.

If a sentient being be in the neighbourhood of such a displacement of the object, his impressions will be modified; but by moving in a suitable manner, he may reconstruct them. For this purpose, all that is required is that the aggregate of the sentient being and the object, considered as forming a single body, shall experience one of those special displacements which I have just called non-Euclidean. This is possible if we suppose that the limbs of these beings dilate according to the same laws as the other bodies of the world they inhabit.

Although from the point of view of our ordinary geometry there is a deformation of the bodies in this displacement, and although their different parts are no longer in the same relative position, nevertheless we shall see that the impressions of the sentient being remain the same as before; in fact, though the mutual distances of the different parts have varied, yet the parts which at first were in contact are still in contact. It follows that tactile impressions will be unchanged. On the other hand, from the hypothesis as to refraction and the curvature of the rays of light, visual impressions will also be unchanged. These imaginary beings will therefore be led to classify the phenomena they observe, and to distinguish among them the "changes of position," which may be corrected by a voluntary correlative movement, just as we do.

If they construct a geometry, it will not be like ours, which is the study of the movements of our invariable solids; it will be the study of the changes of position which they will have thus distinguished, and will be "non-Euclidean displacements," and *this will be non-Euclidean geometry*. So that beings like ourselves, educated in such a world, will not have the same geometry as ours.

The World of Four Dimensions. — Just as we have pictured to ourselves a non-Euclidean world, so we may picture a world of four dimensions.

The sense of light, even with one eye, together with the muscular sensations relative to the movements of the eyeball, will suffice to enable us to conceive of space of three dimensions. The images of external objects are painted on the retina, which is a plane of two dimensions; these are *perspectives*. But as eye and objects are movable, we see in succession different perspectives of the same body taken from different points of view. We find at the same time that the transition from one perspective to another is often accompanied by muscular sensations. If the transition from the perspective A to the perspective B, and that of the perspective A' to the perspective B' are accompanied by the same muscular sensations, we connect them as we do other operations of the same nature. Then when we study the laws according to which these operations are combined, we see that they form a group, which has the same structure as that of the movements of invariable solids. Now, we have seen that it is from the properties of this group that we derive the idea of geometrical space and that of three dimensions. We thus understand how these perspectives gave rise to the conception of three dimensions, although each perspective is of only two dimensions, — because *they succeed each other according to certain laws*. Well, in the same way that we draw the perspective of a three-dimensional figure on a plane, so we can draw that of a four-dimensional figure on a canvas of three (or two) dimensions. To a geometer this is but child's play. We can even draw several perspectives of the same figure from several different points of view. We can easily represent to ourselves these perspectives, since they are of only three dimensions. Imagine that the different perspectives of one and the same object to occur in succession, and that the transition from one to the other is accompanied by muscular sensations. It is understood that we shall consider two of these transitions as two operations of the same nature when they are associated with the same muscular sensations. There is nothing, then, to prevent us from imagining that these operations are combined according to any law we choose — for instance, by forming a group with the same structure as that of the movements of an invariable four-dimensional solid. In this there is nothing that we cannot represent to ourselves, and, moreover, these sensations are those which a being would experience who has a retina of two dimensions, and who may be displaced in space of four dimensions. In this sense we may say that we can represent to ourselves the fourth dimension.

Conclusions. — It is seen that experiment plays a considerable rôle in the genesis of geometry; but it would be a mistake to conclude from that that geometry is, even in part, an experimental science. If it were experimental, it would only be approximative and provisory. And what a rough approximation it would be! Geometry would be only the study of the movements of solid bodies; but, in reality, it is not concerned with natural solids: its object is certain ideal solids, absolutely invariable, which are but a greatly simplified and very remote image of them. The concept of these ideal bodies is entirely mental, and experiment is but the opportunity which enables us to reach the idea. The object of geometry is the study of a particular "group"; but the general concept of group pre-exists in our minds, at least potentially. It is imposed on us not as a form of our sensitiveness, but as a form of our understanding; only, from among all possible groups, we must choose one that will be the *standard*, so to speak, to which we shall refer natural phenomena.

Experiment guides us in this choice, which it does not impose on us. It tells us not what is the truest, but what is the most convenient geometry. It will be noticed that my description of these fantastic worlds has required no language other than that of ordinary geometry. Then, were we transported to those worlds, there would be no need to change that language. Beings educated there would no doubt find it more convenient to create a geometry different from ours, and better adapted to their impressions; but as for us, in the presence of the same impressions, it is certain that we should not find it more convenient to make a change.

CHAPTER V.
EXPERIMENT AND GEOMETRY.

1. I have on several occasions in the preceding pages tried to show how the principles of geometry are not experimental facts, and that in particular Euclid's postulate cannot be proved by experiment. However convincing the reasons already given may appear to me, I feel I must dwell upon them, because there is a profoundly false conception deeply rooted in many minds.

2. Think of a material circle, measure its radius and circumference, and see if the ratio of the two lengths is equal to π. What have we done? We have made an experiment on the properties of the matter with which this *roundness* has been realised, and of which the measure we used is made.

3. *Geometry and Astronomy.* — The same question may also be asked in another way. If Lobatschewsky's geometry is true, the parallax of a very distant star will be finite. If Riemann's is true, it will be negative. These are the results which seem within the reach of experiment, and it is hoped that astronomical observations may enable us to decide between the two geometries. But what we call a straight line in astronomy is simply the path of a ray of light. If, therefore, we were to discover negative parallaxes, or to prove that all parallaxes are higher than a certain limit, we should have a choice between two conclusions: we could give up Euclidean geometry, or modify the laws of optics, and suppose that light is not rigorously propagated in a straight line. It is needless to add that every one would look upon this solution as the more advantageous. Euclidean geometry, therefore, has nothing to fear from fresh experiments.

4. Can we maintain that certain phenomena which are possible in Euclidean space would be impossible in non-Euclidean space, so that experiment in establishing these phenomena would directly contradict the non-Euclidean hypothesis? I think that such a question cannot be seriously asked. To me it is exactly equivalent to the following, the absurdity of which is obvious: — There are lengths which can be expressed in metres and centimetres, but cannot be measured in toises, feet, and inches; so that experiment, by ascertaining the existence of these lengths, would directly contradict this hypothesis, that there are toises divided into six feet. Let us look at the question a little more closely. I assume that the straight line in Euclidean space possesses any two properties, which I shall call A and B; that in non-Euclidean space it still possesses the property A, but no longer possesses the property B; and, finally, I assume that in both Euclidean and non-Euclidean space the straight line is the only line that possesses the property A. If this were so, experiment would be able to decide between the hypotheses of Euclid and Lobatschewsky. It would be found that some concrete object, upon which we can experiment — for example, a pencil of rays of light — possesses the property A. We should conclude that it is rectilinear, and we should then endeavour to find out if it does, or does not, possess the property B. But *it is not so*. There exists no property which can, like this property A, be an absolute criterion enabling us to recognise the straight line, and to distinguish it from every other line. Shall we say, for instance, "This property will be the following: the straight line is a line such that a figure of which this line is a part can move without the mutual distances of its points varying, and in such a way that all the points in this straight line remain fixed"? Now, this is a property which in either Euclidean or non-Euclidean space belongs to the straight line, and belongs to it alone. But how can we ascertain by experiment if it belongs to any particular concrete object? Distances must be measured, and how shall we know that any concrete magnitude which I have

measured with my material instrument really represents the abstract distance? We have only removed the difficulty a little farther off. In reality, the property that I have just enunciated is not a property of the straight line alone; it is a property of the straight line and of distance. For it to serve as an absolute criterion, we must be able to show, not only that it does not also belong to any other line than the straight line and to distance, but also that it does not belong to any other line than the straight line, and to any other magnitude than distance. Now, that is not true, and if we are not convinced by these considerations, I challenge any one to give me a concrete experiment which can be interpreted in the Euclidean system, and which cannot be interpreted in the system of Lobatschewsky. As I am well aware that this challenge will never be accepted, I may conclude that no experiment will ever be in contradiction with Euclid's postulate; but, on the other hand, no experiment will ever be in contradiction with Lobatschewsky's postulate.

5. But it is not sufficient that the Euclidean (or non-Euclidean) geometry can ever be directly contradicted by experiment. Nor could it happen that it can only agree with experiment by a violation of the principle of sufficient reason, and of that of the relativity of space. Let me explain myself. Consider any material system whatever. We have to consider on the one hand the "state" of the various bodies of this system — for example, their temperature, their electric potential, etc.; and on the other hand their position in space. And among the data which enable us to define this position we distinguish the mutual distances of these bodies that define their relative positions, and the conditions which define the absolute position of the system and its absolute orientation in space. The law of the phenomena which will be produced in this system will depend on the state of these bodies, and on their mutual distances; but because of the relativity and the inertia of space, they will not depend on the absolute position and orientation of the system. In other words, the state of the bodies and their mutual distances at any moment will solely depend on the state of the same bodies and on their mutual distances at the initial moment, but will in no way depend on the absolute initial position of the system and of its absolute initial orientation. This is what we shall call, for the sake of abbreviation, *the law of relativity*.

So far I have spoken as a Euclidean geometer. But I have said that an experiment, whatever it may be, requires an interpretation on the Euclidean hypothesis; it equally requires one on the non-Euclidean hypothesis. Well, we have made a series of experiments. We have interpreted them on the Euclidean hypothesis, and we have recognised that these experiments thus interpreted do not violate this "law of relativity." We now interpret them on the non-Euclidean hypothesis. This is always possible, only the non-Euclidean distances of our different bodies in this new interpretation will not generally be the same as the Euclidean distances in the primitive interpretation. Will our experiment interpreted in this new manner be still in agreement with our "law of relativity," and if this agreement had not taken place, would we not still have the right to say that experiment has proved the falsity of non-Euclidean geometry? It is easy to see that this is an idle fear. In fact, to apply the law of relativity in all its rigour, it must be applied to the entire universe; for if we were to consider only a part of the universe, and if the absolute position of this part were to vary, the distances of the other bodies of the universe would equally vary; their influence on the part of the universe considered might therefore increase or diminish, and this might modify the laws of the phenomena which take place in it. But if our system is the entire universe, experiment is powerless to give us any opinion on its position and its absolute orientation in space. All that our instruments, however perfect they may be, can let us know will be the state of the different parts of the universe, and their mutual distances. Hence, our law of relativity may be enunciated as follows: — The readings that we can make with our instruments at any given moment will depend only on the readings that we were able to make on the same instruments at the initial moment. Now such an enunciation is independent of all interpretation by experiments. If the law is true in the Euclidean interpretation, it will be also true in the non-

Euclidean interpretation. Allow me to make a short digression on this point. I have spoken above of the data which define the position of the different bodies of the system. I might also have spoken of those which define their velocities. I should then have to distinguish the velocity with which the mutual distances of the different bodies are changing, and on the other hand the velocities of translation and rotation of the system; that is to say, the velocities with which its absolute position and orientation are changing. For the mind to be fully satisfied, the law of relativity would have to be enunciated as follows: — The state of bodies and their mutual distances at any given moment, as well as the velocities with which those distances are changing at that moment, will depend only on the state of those bodies, on their mutual distances at the initial moment, and on the velocities with which those distances were changing at the initial moment. But they will not depend on the absolute initial position of the system nor on its absolute orientation, nor on the velocities with which that absolute position and orientation were changing at the initial moment. Unfortunately, the law thus enunciated does not agree with experiments — at least, as they are ordinarily interpreted. Suppose a man were translated to a planet, the sky of which was constantly covered with a thick curtain of clouds, so that he could never see the other stars. On that planet he would live as if it were isolated in space. But he would notice that it revolves, either by measuring its ellipticity (which is ordinarily done by means of astronomical observations, but which could be done by purely geodesic means), or by repeating the experiment of Foucault's pendulum. The absolute rotation of this planet might be clearly shown in this way. Now, here is a fact which shocks the philosopher, but which the physicist is compelled to accept. We know that from this fact Newton concluded the existence of absolute space. I myself cannot accept this way of looking at it. I shall explain why in Part III., but for the moment it is not my intention to discuss this difficulty. I must therefore resign myself, in the enunciation of the law of relativity, to including velocities of every kind among the data which define the state of the bodies. However that may be, the difficulty is the same for both Euclid's geometry and for Lobatschewsky's. I need not therefore trouble about it further, and I have only mentioned it incidentally. To sum up, whichever way we look at it, it is impossible to discover in geometric empiricism a rational meaning.

6. Experiments only teach us the relations of bodies to one another. They do not and cannot give us the relations of bodies and space, nor the mutual relations of the different parts of space. "Yes!" you reply, "a single experiment is not enough, because it only gives us one equation with several unknowns; but when I have made enough experiments I shall have enough equations to calculate all my unknowns." If I know the height of the main-mast, that is not sufficient to enable me to calculate the age of the captain. When you have measured every fragment of wood in a ship you will have many equations, but you will be no nearer knowing the captain's age. All your measurements bearing on your fragments of wood can tell you only what concerns those fragments; and similarly, your experiments, however numerous they may be, referring only to the relations of bodies with one another, will tell you nothing about the mutual relations of the different parts of space.

7. Will you say that if the experiments have reference to the bodies, they at least have reference to the geometrical properties of the bodies. First, what do you understand by the geometrical properties of bodies? I assume that it is a question of the relations of the bodies to space. These properties therefore are not reached by experiments which only have reference to the relations of bodies to one another, and that is enough to show that it is not of those properties that there can be a question. Let us therefore begin by making ourselves clear as to the sense of the phrase: geometrical properties of bodies. When I say that a body is composed of several parts, I presume that I am thus enunciating a geometrical property, and that will be true even if I agree to give the improper name of points to the very small parts I am considering. When I say that this or that part of a certain body is in contact with this or that part of another body, I am enunciating a proposition which concerns the mutual relations of the two bodies, and not their relations with space. I assume that you will agree with me that these

are not geometrical properties. I am sure that at least you will grant that these properties are independent of all knowledge of metrical geometry. Admitting this, I suppose that we have a solid body formed of eight thin iron rods, *oa, ob, oc, od, oe, of, og, oh*, connected at one of their extremities, *o*. And let us take a second solid body — for example, a piece of wood, on which are marked three little spots of ink which I shall call *α β γ*. I now suppose that we find that we can bring into contact *α β γ* with *ago*; by that I mean *α* with *a*, and at the same time *β* with *g*, and *γ* with *o*. Then we can successively bring into contact *α β γ* with *bgo, cgo, dgo, ego, fgo*, then with *aho, bho, cho, dho, eho, fho*; and then *α γ* successively with *ab, bc, cd, de, ef, fa*. Now these are observations that can be made without having any idea beforehand as to the form or the metrical properties of space. They have no reference whatever to the "geometrical properties of bodies." These observations will not be possible if the bodies on which we experiment move in a group having the same structure as the Lobatschewskian group (I mean according to the same laws as solid bodies in Lobatschewsky's geometry). They therefore suffice to prove that these bodies move according to the Euclidean group; or at least that they do not move according to the Lobatschewskian group. That they may be compatible with the Euclidean group is easily seen; for we might make them so if the body *α β γ* were an invariable solid of our ordinary geometry in the shape of a right-angled triangle, and if the points *abcdefgh* were the vertices of a polyhedron formed of two regular hexagonal pyramids of our ordinary geometry having *abcdef* as their common base, and having the one *g* and the other *h* as their vertices. Suppose now, instead of the previous observations, we note that we can as before apply *α β γ* successively to *ago, bgo, cgo, dgo, ego, fgo, aho, bho, cho, dho, eho, fho*, and then that we can apply *αβ* (and no longer *α γ*) successively to *ab, bc, cd, de, ef*, and *fa*. These are observations that could be made if non-Euclidean geometry were true. If the bodies *α β γ, oabcdefgh* were invariable solids, if the former were a right-angled triangle, and the latter a double regular hexagonal pyramid of suitable dimensions. These new verifications are therefore impossible if the bodies move according to the Euclidean group; but they become possible if we suppose the bodies to move according to the Lobatschewskian group. They would therefore suffice to show, if we carried them out, that the bodies in question do not move according to the Euclidean group. And so, without making any hypothesis on the form and the nature of space, on the relations of the bodies and space, and without attributing to bodies any geometrical property, I have made observations which have enabled me to show in one case that the bodies experimented upon move according to a group, the structure of which is Euclidean, and in the other case, that they move in a group, the structure of which is Lobatschewskian. It cannot be said that all the first observations would constitute an experiment proving that space is Euclidean, and the second an experiment proving that space is non-Euclidean; in fact, it might be imagined (note that I use the word *imagined*) that there are bodies moving in such a manner as to render possible the second series of observations: and the proof is that the first mechanic who came our way could construct it if he would only take the trouble. But you must not conclude, however, that space is non-Euclidean. In the same way, just as ordinary solid bodies would continue to exist when the mechanic had constructed the strange bodies I have just mentioned, he would have to conclude that space is both Euclidean and non-Euclidean. Suppose, for instance, that we have a large sphere of radius *R*, and that its temperature decreases from the centre to the surface of the sphere according to the law of which I spoke when I was describing the non-Euclidean world. We might have bodies whose dilatation is negligeable, and which would behave as ordinary invariable solids; and, on the other hand, we might have very dilatable bodies, which would behave as non-Euclidean solids. We might have two double pyramids *oabcdefgh* and *o'a'b'c'd'e'f'g'h'*, and two triangles *α β γ* and *α' β' γ'*. The first double pyramid would be rectilinear, and the second curvilinear. The triangle *α β γ* would consist of undilatable matter, and the other of very dilatable matter. We might therefore make our first observations with the double pyramid *o'a'h'* and the triangle *α' β' γ'*.

And then the experiment would seem to show — first, that Euclidean geometry is true, and then that it is false. Hence, *experiments have reference not to space but to bodies.*

SUPPLEMENT.

8. To round the matter off, I ought to speak of a very delicate question, which will require considerable development; but I shall confine myself to summing up what I have written in the *Revue de Métaphysique et de Morale* and in the *Monist.* When we say that space has three dimensions, what do we mean? We have seen the importance of these "internal changes" which are revealed to us by our muscular sensations. They may serve to characterise the different attitudes of our body. Let us take arbitrarily as our origin one of these attitudes, A. When we pass from this initial attitude to another attitude B we experience a series of muscular sensations, and this series S of muscular sensations will define B. Observe, however, that we shall often look upon two series S and S' as defining the same attitude B (since the initial and final attitudes A and B remaining the same, the intermediary attitudes of the corresponding sensations may differ). How then can we recognise the equivalence of these two series? Because they may serve to compensate for the same external change, or more generally, because, when it is a question of compensation for an external change, one of the series may be replaced by the other. Among these series we have distinguished those which can alone compensate for an external change, and which we have called "displacements." As we cannot distinguish two displacements which are very close together, the aggregate of these displacements presents the characteristics of a physical continuum. Experience teaches us that they are the characteristics of a physical continuum of six dimensions; but we do not know as yet how many dimensions space itself possesses, so we must first of all answer another question. What is a point in space? Every one thinks he knows, but that is an illusion. What we see when we try to represent to ourselves a point in space is a black spot on white paper, a spot of chalk on a blackboard, always an object. The question should therefore be understood as follows: — What do I mean when I say the object B is at the point which a moment before was occupied by the object A? Again, what criterion will enable me to recognise it? I mean that *although I have not moved* (my muscular sense tells me this), my finger, which just now touched the object A, is now touching the object B. I might have used other criteria — for instance, another finger or the sense of sight — but the first criterion is sufficient. I know that if it answers in the affirmative all other criteria will give the same answer. I know it from experiment. I cannot know it *à priori.* For the same reason I say that touch cannot be exercised at a distance; that is another way of enunciating the same experimental fact. If I say, on the contrary, that sight is exercised at a distance, it means that the criterion furnished by sight may give an affirmative answer while the others reply in the negative.

To sum up. For each attitude of my body my finger determines a point, and it is that and that only which defines a point in space. To each attitude corresponds in this way a point. But it often happens that the same point corresponds to several different attitudes (in this case we say that our finger has not moved, but the rest of our body has). We distinguish, therefore, among changes of attitude those in which the finger does not move. How are we led to this? It is because we often remark that in these changes the object which is in touch with the finger remains in contact with it. Let us arrange then in the same class all the attitudes which are deduced one from the other by one of the changes that we have thus distinguished. To all these attitudes of the same class will correspond the same point in space. Then to each class will correspond a point, and to each point a class. Yet it may be said that what we get from this experiment is not the point, but the class of changes, or, better still, the corresponding class of muscular sensations. Thus, when we say that space has three dimensions, we merely mean that the aggregate of these classes appears to us with the characteristics of a physical continuum of three dimensions. Then if, instead of defining the points in space with the aid of the first finger, I use, for example, another finger, would the results be the same? That is by no means à priori evident. But, as we have seen, experiment has shown us that all our criteria are in agreement, and this enables us to answer in the affirmative. If we recur to what we have called displacements, the aggregate of which forms, as we have seen, a group, we shall be brought to distinguish those in which a finger does not move; and by what has preceded, those are the displacements which characterise a point in space, and their aggregate will form a sub-group of our group. To each sub-group of this kind, then, will correspond a point in space. We might be tempted to conclude that experiment has taught us the number of dimensions of space; but in reality our experiments have referred not to space, but to our body and its relations with neighbouring objects. What is more, our experiments are exceeding crude. In our mind the latent idea of a certain number of groups pre-existed; these are the groups with which Lie's theory is concerned. Which shall we choose to form a kind of standard by which to compare natural phenomena? And when this group is chosen, which of the sub-groups shall we take to characterise a point in space? Experiment has guided us by showing us what choice adapts itself best to the properties of our body; but there its rôle ends.

PART III.

FORCE.

CHAPTER VI.
THE CLASSICAL MECHANICS.

THE English teach mechanics as an experimental science; on the Continent it is taught always more or less as a deductive and *à priori* science. The English are right, no doubt. How is it that the other method has been persisted in for so long; how is it that Continental scientists who have tried to escape from the practice of their predecessors have in most cases been unsuccessful? On the other hand, if the principles of mechanics are only of experimental origin, are they not merely approximate and provisory? May we not be some day compelled by new experiments to modify or even to abandon them? These are the questions which naturally arise, and the difficulty of solution is largely due to the fact that treatises on mechanics do not clearly distinguish between what is experiment, what is mathematical reasoning, what is convention, and what is hypothesis. This is not all.

1. There is no absolute space, and we only conceive of relative motion; and yet in most cases mechanical facts are enunciated as if there is an absolute space to which they can be referred.

2. There is no absolute time. When we say that two periods are equal, the statement has no meaning, and can only acquire a meaning by a convention.

3. Not only have we no direct intuition of the equality of two periods, but we have not even direct intuition of the simultaneity of two events occurring in two different places. I have explained this in an article entitled "**Mesure du Temps**."[1]

4. Finally, is not our Euclidean geometry in itself only a kind of convention of language? Mechanical facts might be enunciated with reference to a non-Euclidean space which would be less convenient but quite as legitimate as our ordinary space; the enunciation would become more complicated, but it still would be possible.

Thus, absolute space, absolute time, and even geometry are not conditions which are imposed on mechanics. All these things no more existed before mechanics than the French language can be logically said to have existed before the truths which are expressed in French. We might endeavour to enunciate the fundamental law of mechanics in a language independent of all these conventions; and no doubt we should in this way get a clearer idea of those laws in themselves. This is what M. Andrade has tried to do, to some extent at any rate, in his *Leçons de Mécanique physique*. Of course the enunciation of these laws would become much more complicated, because all these conventions have been adopted for the very purpose of abbreviating and simplifying the enunciation. As far as we are concerned, I shall ignore all these difficulties; not because I disregard them, far from it; but because they have received sufficient attention in the first two parts of the book. Provisionally, then, we shall admit absolute time and Euclidean geometry.

The Principle of Inertia. — A body under the action of no force can only move uniformly in a straight line. Is this a truth imposed on the mind *à priori*? If this be so, how is it that the Greeks ignored it? How could they have believed that motion ceases with the cause of motion? or, again, that every body, if there is nothing to prevent it, will move in a circle, the noblest of all forms of motion?

If it be said that the velocity of a body cannot change, if there is no reason for it to change, may we not just as legitimately maintain that the position of a body cannot change, or that the curvature of its path cannot change, without the agency of an external cause? Is, then, the principle of inertia, which is not an à priori truth, an experimental fact? Have there ever been experiments on bodies acted on by no forces? and, if so, how did we know that no forces were acting? The usual instance is that of a ball rolling for a very long time on a marble table; but why do we say it is under the action of no force? Is it because it is too remote from all other bodies to experience any sensible action? It is not further from the earth than if it were thrown freely into the air; and we all know that in that case it would be subject to the attraction of the earth. Teachers of mechanics usually pass rapidly over the example of the ball, but they add that the principle of inertia is verified indirectly by its consequences. This is very badly expressed; they evidently mean that various consequences may be verified by a more general principle, of which the principle of inertia is only a particular case. I shall propose for this general principle the following enunciation: — The acceleration of a body depends only on its position and that of neighbouring bodies, and on their velocities. Mathematicians would say that the movements of all the material molecules of the universe depend on differential equations of the second order. To make it clear that this is really a generalisation of the law of inertia we may again have recourse to our imagination. The law of inertia, as I have said above, is not imposed on us à priori; other laws would be just as compatible with the principle of sufficient reason. If a body is not acted upon by a force, instead of supposing that its velocity is unchanged we may suppose that its position or its acceleration is unchanged.

Let us for a moment suppose that one of these two laws is a law of nature, and substitute it for the law of inertia: what will be the natural generalisation? A moment's reflection will show us. In the first case, we may suppose that the velocity of a body depends only on its position and that of neighbouring bodies; in the second case, that the variation of the acceleration of a body depends only on the position of the body and of neighbouring bodies, on their velocities and accelerations; or, in mathematical terms, the differential equations of the motion would be of the first order in the first case and of the third order in the second.

Let us now modify our supposition a little. Suppose a world analogous to our solar system, but one in which by a singular chance the orbits of all the planets have neither eccentricity nor inclination; and further, I suppose that the masses of the planets are too small for their mutual perturbations to be sensible. Astronomers living in one of these planets would not hesitate to conclude that the orbit of a star can only be circular and parallel to a certain plane; the position of a star at a given moment would then be sufficient to determine its velocity and path. The law of inertia which they would adopt would be the former of the two hypothetical laws I have mentioned.

Now, imagine this system to be some day crossed by a body of vast mass and immense velocity coming from distant constellations. All the orbits would be profoundly disturbed. Our astronomers would not be greatly astonished. They would guess that this new star is in itself quite capable of doing all the mischief; but, they would say, as soon as it has passed by, order will again be established. No doubt the distances of the planets from the sun will not be the same as before the cataclysm, but the orbits will become circular again as soon as the disturbing cause has disappeared. It would be only when the perturbing body is remote, and when the orbits, instead of being circular are found to be elliptical, that the astronomers would find out their mistake, and discover the necessity of reconstructing their mechanics.

I have dwelt on these hypotheses, for it seems to me that we can clearly understand our generalised law of inertia only by opposing it to a contrary hypothesis.

Has this generalised law of inertia been verified by experiment, and can it be so verified? When Newton wrote the Principia, he certainly regarded this truth as experimentally acquired and demonstrated. It was so in his eyes, not only from the anthropomorphic conception to which I shall later refer, but also because of the work of Galileo. It was so proved by the laws of Kepler. According to those laws, in fact, the path of a planet is entirely determined by its initial position and initial velocity; this, indeed, is what our generalised law of inertia requires.

For this principle to be only true in appearance — lest we should fear that some day it must be replaced by one of the analogous principles which I opposed to it just now — we must have been led astray by some amazing chance such as that which had led into error our imaginary astronomers. Such an hypothesis is so unlikely that it need not delay us. No one will believe that there can be such chances; no doubt the probability that two eccentricities are both exactly zero is not smaller than the probability that one is 0.1 and the other 0.2. The probability of a simple event is not smaller than that of a complex one. If, however, the former does occur, we shall not attribute its occurrence to chance; we shall not be inclined to believe that nature has done it deliberately to deceive us. The hypothesis of an error of this kind being discarded, we may admit that so far as astronomy is concerned our law has been verified by experiment.

But Astronomy is not the whole of Physics. May we not fear that some day a new experiment will falsify the law in some domain of physics? An experimental law is always subject to revision; we may always expect to see it replaced by some other and more exact law. But no one seriously thinks that the law of which we speak will ever be abandoned or amended. Why? Precisely because it will never be submitted to a decisive test.

In the first place, for this test to be complete, all the bodies of the universe must return with their initial velocities to their initial positions after a certain time. We ought then to find that they would resume their original paths. But this test is impossible; it can be only partially applied, and even when it is applied there will still be some bodies which will not return to their original positions. Thus there will be a ready explanation of any breaking down of the law.

Yet this is not all. In Astronomy we *see* the bodies whose motion we are studying, and in most cases we grant that they are not subject to the action of other invisible bodies. Under these conditions, our law must certainly be either verified or not. But it is not so in Physics. If physical phenomena are due to motion, it is to the motion of molecules which we cannot see. If, then, the acceleration of bodies we cannot see depends on something else than the positions or velocities of other visible bodies or of invisible molecules, the existence of which we have been led previously to admit, there is nothing to prevent us from supposing that this something else is the position or velocity of other molecules of which we have not so far suspected the existence. The law will be safeguarded. Let me express the same thought in another form in mathematical language. Suppose we are observing n molecules, and find that their $3n$ co-ordinates satisfy a system of $3n$ differential equations of the fourth order (and not of the second, as required by the law of inertia). We know that by introducing $3n$ variable auxiliaries, a system of $3n$ equations of the fourth order may be reduced to a system of $6n$ equations of the second order. If, then, we suppose that the $3n$ auxiliary variables represent the co-ordinates of n invisible molecules, the result is again conform able to the law of inertia. To sum up, this law, verified experimentally in some particular cases, may be extended fearlessly to the most general cases; for we know that in these general cases it can neither be confirmed nor contradicted by experiment.

The Law of Acceleration. — The acceleration of a body is equal to the force which acts on it divided by its mass.

Can this law be verified by experiment? If so, we have to measure the three magnitudes mentioned in the enunciation: acceleration, force, and mass. I admit that acceleration may be measured, because I pass over the difficulty arising from the measurement of time. But how are we to measure force and mass? We do not even know what they are. What is mass? Newton replies: "The product of the volume and the density." "It were better to say," answer Thomson and Tait, "that density is the quotient of the mass by the volume." What is force? "It is," replies Lagrange, "that which moves or tends to move a body." "It is," according to Kirchoff, "the product of the mass and the acceleration." Then why not say that mass is the quotient of the force by the acceleration? These difficulties are insurmountable.

When we say force is the cause of motion, we are talking metaphysics; and this definition, if we had to be content with it, would be absolutely fruitless, would lead to absolutely nothing. For a definition to be of any use it must tell us how to measure force; and that is quite sufficient, for it is by no means necessary to tell what force is in itself, nor whether it is the cause or the effect of motion. We must therefore first define what is meant by the equality of two forces. When are two forces equal? We are told that it is when they give the same acceleration to the same mass, or when acting in opposite directions they are in equilibrium. This definition is a sham. A force applied to a body cannot be uncoupled and applied to another body as an engine is uncoupled from one train and coupled to another. It is therefore impossible to say what acceleration such a force, applied to such a body, would give to another body if it were applied to it. It is impossible to tell how two forces which are not acting in exactly opposite directions would be have if they were acting in opposite directions. It is this definition which we try to materialise, as it were, when we measure a force with a dynamometer or with a balance. Two forces, F and F', which I suppose, for simplicity, to be acting vertically upwards, are respectively applied to two bodies, C and C'. I attach a body weighing P first to C and then to C'; if there is equilibrium in both cases I conclude that the two forces F and F' are equal, for they are both equal to the weight of the body P. But am I certain that the body P has kept its weight when I transferred it from the first body to the second? Far from it. I am certain of the contrary. I know that the magnitude of the weight varies from one point to another, and that it is greater, for instance, at the pole than at the equator. No doubt the difference is very small, and we neglect it in practice; but a definition must have mathematical rigour; this rigour does not exist. What I say of weight would apply equally to the force of the spring of a dynamometer, which would vary according to temperature and many other circumstances. Nor is this all. We cannot say that the weight of the body P is applied to the body C and keeps in equilibrium the force F. What is applied to the body C is the action of the body P on the body C. On the other hand, the body P is acted on by its weight, and by the reaction R of the body C on P the forces F and A are equal, because they are in equilibrium; the forces A and R are equal by virtue of the principle of action and reaction; and finally, the force R and the weight P are equal because they are in equilibrium. From these three equalities we deduce the equality of the weight P and the force F.

Thus we are compelled to bring into our definition of the equality of two forces the principle of the equality of action and reaction; *hence this principle can no longer be regarded as an experimental law but only as a definition.*

To recognise the equality of two forces we are then in possession of two rules: the equality of two forces in equilibrium and the equality of action and reaction. But, as we have seen, these are not sufficient, and we are compelled to have recourse to a third rule, and to admit that certain forces — the weight of a body, for instance — are constant in magnitude and direction. But this third rule is an experimental law. It is only approximately true: *it is a bad definition.* We are therefore reduced to Kirchoff's definition: force is the product of the mass and the acceleration. This law of Newton in its turn ceases to be regarded as an experimental law, it is now only a definition. But as a definition it is insufficient, for we do not know what mass is. It enables us, no doubt, to calculate the ratio of two forces applied at different times to the same body, but it tells us nothing about the ratio of two forces applied to two different bodies. To fill up the gap we must have recourse to Newton's third law, the equality of action and reaction, still regarded not as an experimental law but as a definition. Two bodies, A and B, act on each other; the acceleration of A, multiplied by the mass of A, is equal to the action of B on A; in the same way the acceleration of B, multiplied by the mass of B, is equal to the reaction of A on B. As, by definition, the action and the reaction are equal, the masses of A and B arc respectively in the inverse ratio of their masses. Thus is the ratio of the two masses defined, and it is for experiment to verify that the ratio is constant.

This would do very well if the two bodies were alone and could be abstracted from the action of the rest of the world; but this is by no means the case. The acceleration of A is not solely due to the action of B, but to that of a multitude of other bodies, C, D, . . . To apply the preceding rule we must decompose the acceleration of A into many components, and find out which of these components is due to the action of B. The decomposition would still be possible if we suppose that the action of C on A is simply added to that of B on A, and that the presence of the body C does not in any way modify the action of B on A, or that the presence of B does not modify the action of C on A; that is, if we admit that any two bodies attract each other, that their mutual action is along their join, and is only dependent on their distance apart; if, in a word, we admit the *hypothesis of central forces*.

We know that to determine the masses of the heavenly bodies we adopt quite a different principle. The law of gravitation teaches us that the attraction of two bodies is proportional to their masses; if r is their distance apart, m and m' their masses, k a constant, then their attraction will be kmm'/r^2. What we are measuring is therefore not mass, the ratio of the force to the acceleration, but the attracting mass; not the inertia of the body, but its attracting power. It is an indirect process, the use of which is not indispensable theoretically. We might have said that the attraction is inversely proportional to the square of the distance, without being proportional to the product of the masses, that it is equal to f/r^2 and not to kmm'. If it were so, we should nevertheless, by observing the *relative* motion of the celestial bodies, be able to calculate the masses of these bodies.

But have we any right to admit the hypothesis of central forces? Is this hypothesis rigorously accurate? Is it certain that it will never be falsified by experiment? Who will venture to make such an assertion? And if we must abandon this hypothesis, the building which has been so laboriously erected must fall to the ground.

We have no longer any right to speak of the component of the acceleration of A which is due to the action of B. We have no means of distinguishing it from that which is due to the action of C or of any other body. The rule becomes inapplicable in the measurement of masses. What then is left of the principle of the equality of action and reaction? If we reject the hypothesis of central forces this principle must go too; the geometrical resultant of all the forces applied to the different bodies of a system abstracted from all external action will be zero. In other words, *the motion of the centre of gravity of this system will be uniform and in a straight line*. Here would seem to be a means of defining mass. The position of the centre of gravity evidently depends on the values given to the masses; we must select these values so that the motion of the centre of gravity is uniform and rectilinear. This will always be possible if Newton's third law holds good, and it will be in general possible only in one way. But no system exists which is abstracted from all external action; every part of the universe is subject, more or less, to the action of the other parts. *The law of the motion of the centre of gravity is only rigorously true when applied to the whole universe.*

But then, to obtain the values of the masses we must find the motion of the centre of gravity of the universe. The absurdity of this conclusion is obvious; the motion of the centre of gravity of the universe will be for ever to us unknown. Nothing, therefore, is left, and our efforts are fruitless. There is no escape from the following definition, which is only a confession of failure: *Masses are co-efficients which it is found convenient to introduce into calculations.*

We could reconstruct our mechanics by giving to our masses different values. The new mechanics would be in contradiction neither with experiment nor with the general principles of dynamics (the principle of inertia, proportionality of masses and accelerations, equality of action and reaction, uniform motion of the centre of gravity in a straight line, and areas). But the equations of this mechanics *would not be so simple*. Let us clearly understand this. It would be only the first terms which would be less simple — *i.e.*, those we already know through experiment; perhaps the small masses could be slightly altered without the *complete* equations gaining or losing in simplicity.

Hertz has inquired if the principles of mechanics are rigorously true. "In the opinion of many physicists it seems inconceivable that experiment will ever alter the impregnable principles of mechanics; and yet, what is due to experiment may always be rectified by experiment." From what we have just seen these fears would appear to be groundless. The principles of dynamics appeared to us first as experimental truths, but we have been compelled to use them as definitions. It is *by definition* that force is equal to the product of the mass and the acceleration; this is a principle which is henceforth beyond the reach of any future experiment. Thus it is by definition that action and reaction are equal and opposite. But then it will be said, these unverifiable principles are absolutely devoid of any significance. They cannot be disproved by experiment, but we can learn from them nothing of any use to us; what then is the use of studying dynamics? This somewhat rapid condemnation would be rather unfair. There is not in Nature any system *perfectly* isolated, perfectly abstracted from all external action; but there are systems which are *nearly* isolated. If we observe such a system, we can study not only the relative motion of its different parts with respect to each other, but the motion of its centre of gravity with respect to the other parts of the universe. We then find that the motion of its centre of gravity is *nearly* uniform and rectilinear in conformity with Newton's Third Law. This is an experimental fact, which cannot be invalidated by a more accurate experiment. What, in fact, would a more accurate experiment teach us? It would teach us that the law is only approximately true, and we know that already. *Thus is explained how experiment may serve as a basis for the principles of mechanics, and yet will never invalidate them.*

Anthropomorphic Mechanics. — It will be said that Kirchoff has only followed the general tendency of mathematicians towards nominalism; from this his skill as a physicist has not saved him. He wanted a definition of a force, and he took the first that came handy; but we do not require a definition of force; the idea of force is primitive, irreducible, indefinable; we all know what it is; of it we have direct intuition. This direct intuition arises from the idea of effort which is familiar to us from childhood. But in the first place, even if this direct intuition made known to us the real nature of force in itself, it would prove to be an insufficient basis for mechanics; it would, moreover, be quite useless. The important thing is not to know what force is, but how to measure it. Everything which does not teach us how to measure it is as useless to the mechanician as, for instance, the subjective idea of heat and cold to the student of heat. This subjective idea cannot be translated into numbers, and is therefore useless; a scientist whose skin is an absolutely bad conductor of heat, and who, therefore, has never felt the sensation of heat or cold, would read a thermometer in just the same way as any one else, and would have enough material to construct the whole of the theory of heat.

Now this immediate notion of effort is of no use to us in the measurement of force. It is clear, for example, that I shall experience more fatigue in lifting a weight of 100 lb. than a man who is accustomed to lifting heavy burdens. But there is more than this. This notion of effort does not teach us the nature of force; it is definitively reduced to a recollection of muscular sensations, and no one will maintain that the sun experiences a muscular sensation when it attracts the earth. All that we can expect to find from it is a symbol, less precise and less convenient than the arrows (to denote direction) used by geometers, and quite as remote from reality.

Anthropomorphism plays a considerable historic rôle in the genesis of mechanics; perhaps it may yet furnish us with a symbol which some minds may find convenient; but it can be the foundation of nothing of a really scientific or philosophical character.

The Thread School. — M. Andrade, in his *Leçons de Mecanique physique*, has modernised anthropomorphic mechanics. To the school of mechanics with which Kirchoff is identified, he opposes a school which is quaintly called the "Thread School."

This school tries to reduce everything to the consideration of certain material systems of negligible mass, regarded in a state of tension and capable of transmitting considerable effort to distant bodies — systems of which the ideal type is the fine string, wire, or *thread*. A thread which transmits any force is slightly lengthened in the direction of that force; the direction of the thread tells us the direction of the force, and the magnitude of the force is measured by the lengthening of the thread.

We may imagine such an experiment as the following: — A body A is attached to a thread; at the other extremity of the thread acts a force which is made to vary until the length of the thread is increased by α, and the acceleration of the body A is recorded. A is then detached, and a body B is attached to the same thread, and the same or another force is made to act until the increment of length again is α, and the acceleration of B is noted. The experiment is then renewed with both A and B until the increment of length is β. The four accelerations observed should be proportional. Here we have an experimental verification of the law of acceleration enunciated above. Again, we may consider a body under the action of several threads in equal tension, and by experiment we determine the direction of those threads when the body is in equilibrium. This is an experimental verification of the law of the composition of forces. But, as a matter of fact, what have we done? We have defined the force acting on the string by the deformation of the thread, which is reasonable enough; we have then assumed that if a body is attached to this thread, the effort which is transmitted to it by the thread is equal to the action exercised by the body on the thread; in fact, we have used the principle of action and reaction by considering it, not as an experimental truth, but as the very definition of force. This definition is quite as conventional as that of Kirchoff, but it is much less general.

All the forces are not transmitted by the thread (and to compare them they would all have to be transmitted by identical threads). If we even admitted that the earth is attached to the sun by an invisible thread, at any rate it will be agreed that we have no means of measuring the increment of the thread. Nine times out of ten, in consequence, our definition will be in default; no sense of any kind can be attached to it, and we must fall back on that of Kirchoff. Why then go on in this roundabout way? You admit a certain definition of force which has a meaning only in certain particular cases. In those cases you verify by experiment that it leads to the law of acceleration. On the strength of these experiments you then take the law of acceleration as a definition of force in all the other cases.

Would it not be simpler to consider the law of acceleration as a definition in all cases, and to regard the experiments in question, not as verifications of that law, but as verifications of the principle of action and reaction, or as proving the deformations of an elastic body depend only on the forces acting on that body? Without taking into account the fact that the conditions in which your definition could be accepted can only be very imperfectly fulfilled, that a thread is never without mass, that it is never isolated from all other forces than the reaction of the bodies attached to its extremities.

The ideas expounded by M. Andrade are none the less very interesting. If they do not satisfy our logical requirements, they give us a better view of the historical genesis of the fundamental ideas of mechanics. The reflections they suggest show us how the human mind passed from a naive anthropomorphism to the present conception of science.

We see that we end with an experiment which is very particular, and as a matter of fact very crude, and we start with a perfectly general law, perfectly precise, the truth of which we regard as absolute. We have, so to speak, freely conferred this certainty on it by looking upon it as a convention.

Are the laws of acceleration and of the composition of forces only arbitrary conventions? Conventions, yes; arbitrary, no — they would be so if we lost sight of the experiments which led the founders of the science to adopt them, and which, imperfect as they were, were sufficient to justify their adoption. It is well from time to time to let our attention dwell on the experimental origin of these conventions.

Footnotes

1. *Revue de Métaphysique et de Morale*, t. vi., pp. 1-13, January, 1898.

CHAPTER VII.
RELATIVE AND ABSOLUTE MOTION.

The Principle of Relative Motion. — Sometimes endeavours have been made to connect the law of acceleration with a more general principle. The movement of any system whatever ought to obey the same laws, whether it is referred to fixed axes or to the movable axes which are implied in uniform motion in a straight line. This is the principle of relative motion; it is imposed upon us for two reasons: the commonest experiment confirms it; the consideration of the contrary hypothesis is singularly repugnant to the mind.

Let us admit it then, and consider a body under the action of a force. The relative motion of this body with respect to an observer moving with a uniform velocity equal to the initial velocity of the body, should be identical with what would be its absolute motion if it started from rest. We conclude that its acceleration must not depend upon its absolute velocity, and from that we attempt to deduce the complete law of acceleration.

For a long time there have been traces of this proof in the regulations for the degree of B. ès Sc. It is clear that the attempt has failed. The obstacle which prevented us from proving the law of acceleration is that we have no definition of force. This obstacle subsists in its entirety, since the principle invoked has not furnished us with the missing definition. The principle of relative motion is none the less very interesting, and deserves to be considered for its own sake. Let us try to enunciate it in an accurate manner. We have said above that the accelerations of the different bodies which form part of an isolated system only depend on their velocities and their relative positions, and not on their velocities and their absolute positions, provided that the movable axes to which the relative motion is referred move uniformly in a straight line; or, if it is preferred, their accelerations depend only on the differences of their velocities and the differences of their co-ordinates, and not on the absolute values of these velocities and co-ordinates. If this principle is true for relative accelerations, or rather for differences of acceleration, by combining it with the law of reaction we shall deduce that it is true for absolute accelerations. It remains to be seen how we can prove that differences of acceleration depend only on differences of velocities and co-ordinates; or, to speak in mathematical language, that these differences of co-ordinates satisfy differential equations of the second order. Can this proof be deduced from experiment or from *à priori* conditions? Remembering what we have said before, the reader will give his own answer. Thus enunciated, in fact, the principle of relative motion curiously resembles what I called above the generalised principle of inertia; it is not quite the same thing, since it is a question of differences of co-ordinates, and not of the co-ordinates themselves. The new principle teaches us something more than the old, but the same discussion applies to it, and would lead to the same conclusions. We need not recur to it.

Newton's Argument. — Here we find a very important and even slightly disturbing question. I have said that the principle of relative motion was not for us simply a result of experiment; and that *à priori* every contrary hypothesis would be repugnant to the mind. But, then, why is the principle only true if the motion of the movable axes is uniform and in a straight line? It seems that it should be imposed upon us with the same force if the motion is accelerated, or at any rate if it reduces to a uniform rotation. In these two cases, in fact, the principle is not true. I need not dwell on the case in which the motion of the axes is in a straight line and not uniform. The paradox does not bear a moment's examination. If I am in a railway carriage, and if the train, striking against any obstacle whatever, is suddenly stopped, I shall be projected on to the opposite side, although I have not been directly acted upon by any force. There is nothing mysterious in that, and if I have not been subject to the action of any external force, the train has experienced an external impact. There can be nothing paradoxical in the relative motion of two bodies being disturbed when the motion of one or the other is modified by an external cause. Nor need I dwell on the case of relative motion referring to axes which rotate uniformly. If the sky were for ever covered with clouds, and if we had no means of observing the stars, we might, nevertheless, conclude that the earth turns round. We should be warned of this fact by the flattening at the poles, or by the experiment of Foucault's pendulum. And yet, would there in this case be any meaning in saying that the earth turns round? If there is no absolute space, can a thing turn without turning with respect to something; and, on the other hand, how can we admit Newton's conclusion and believe in absolute space? But it is not sufficient to state that all possible solutions are equally unpleasant to us. We must analyse in each case the reason of our dislike, in order to make our choice with the knowledge of the cause. The long discussion which follows must, therefore, be excused.

Let us resume our imaginary story. Thick clouds hide the stars from men who cannot observe them, and even are ignorant of their existence. How will those men know that the earth turns round? No doubt, for a longer period than did our ancestors, they will regard the soil on which they stand as fixed and immovable! They will wait a much longer time than we did for the coming of a Copernicus; but this Copernicus will come at last. How will he come? In the first place, the mechanical school of this world would not run their heads against an absolute contradiction. In the theory of relative motion we observe, besides real forces, two imaginary forces, which we call ordinary centrifugal force and compounded centrifugal force. Our imaginary scientists can thus explain everything by looking upon these two forces as real, and they would not see in this a contradiction of the generalised principle of inertia, for these forces would depend, the one on the relative positions of the different parts of the system, such as real attractions, and the other on their relative velocities, as in the case of real frictions. Many difficulties, however, would before long awaken their attention. If they succeeded in realising an isolated system, the centre of gravity of this system would not have an approximately rectilinear path. They could invoke, to explain this fact, the centrifugal forces which they would regard as real, and which, no doubt, they would attribute to the mutual actions of the bodies — only they would not see these forces vanish at great distances — that is to say, in proportion as the isolation is better realised. Far from it. Centrifugal force increases indefinitely with distance. Already this difficulty would seem to them sufficiently serious, but it would not detain them for long. They would soon imagine some very subtle medium analogous to our ether, in which all bodies would be bathed, and which would exercise on them a repulsive action. But that is not all. Space is symmetrical — yet the laws of motion would present no symmetry. They should be able to distinguish between right and left. They would see, for instance, that cyclones always turn in the same direction, while for reasons of symmetry they should turn indifferently in any direction. If our scientists were able by dint of much hard work to make their universe perfectly symmetrical, this symmetry would not subsist, although there is no apparent reason why it should be disturbed in one direction more than in another. They would extract this from the situation no doubt — they would invent something which would not be more extraordinary than the glass spheres of Ptolemy, and would thus go on accumulating complications until the long-expected Copernicus would sweep them all away with a single blow, saying it is much more simple to admit that the earth turns round. Just as our Copernicus said to us: "It is more convenient to suppose that the earth turns round, because the laws of astronomy are thus expressed in a more simple language," so he would say to them: "It is more convenient to suppose that the earth turns round, because the laws of mechanics are thus expressed in much more simple language. That does not prevent absolute space — that is to say, the point to which we must refer the earth to know if it really does turn round — from having no objective existence. And hence this affirmation: "the earth turns round," has no meaning, since it cannot be verified by experiment; since such an experiment not only cannot be realised or even dreamed of by the most daring Jules Verne, but cannot even be conceived of without contradiction; or, in other words, these two propositions, "the earth turns round," and, "it is more convenient to suppose that the earth turns round," have one and the same meaning. There is nothing more in one than in the other. Perhaps they will not be content with this, and may find it surprising that among all the hypotheses, or rather all the conventions, that can be made on this subject there is one which is more convenient than the rest? But if we have admitted it without difficulty when it is a question of the laws of astronomy, why should we object when it is a question of the laws of mechanics? We have seen that the co-ordinates of bodies are determined by differential equations of the second order, and that so are the differences of these co-ordinates. This is what we have called the generalised principle of inertia, and the principle of relative motion. If the distances of these bodies were determined in the same way by equations of the second order, it seems that the mind should be entirely satisfied. How far does the mind receive this satisfaction, and why is it not content with it? To explain this we had better take a simple example. I assume a system analogous to our solar system, but in which fixed stars foreign to this system cannot be perceived, so that astronomers can only observe the mutual distances of planets and the sun, and not the absolute longitudes of the planets. If we deduce directly from Newton's law the differential equations which define the variation of these distances, these equations will not be of the second order. I mean that if, outside Newton's law, we knew the initial values of these distances and of their derivatives with respect to time — that would not be sufficient to determine the values of these same distances at an ulterior moment. A datum would be still lacking, and this datum might be, for example, what astronomers call the area-constant. But here we may look at it from two different points of view. We may consider two kinds of constants. In the eyes of the physicist the world reduces to a series of phenomena depending, on the one hand, solely on initial phenomena, and, on the other hand, on the laws connecting consequence and antecedent. If observation then teaches us that a certain quantity is a constant, we shall have a choice of two ways of looking at it. So let us admit that there is a law which requires that this quantity shall not vary, but that by chance it has been found to have had in the beginning of time this value rather than that, a value that it has kept ever since. This quantity might then be called an *accidental* constant. Or again, let us admit on the contrary that there is a law of nature which imposes on this quantity this value and not that. We shall then have what may be called an *essential* constant. For example, in virtue of the laws of Newton the duration of the revolution of the earth must be constant. But if it is 366 and something sidereal days, and not 300 or 400, it is because of some initial chance or other. It is an *accidental* constant. If, on the other hand, the exponent of the distance which figures in the expression of the attractive force is equal to -2 and not to -3, it is not by chance, but because it is required by Newton's law. It is an *essential* constant. I do not know if this manner of giving to chance its share is legitimate in itself, and if there is not some artificiality about this distinction; but it is certain at least that in proportion as Nature has secrets, she will be strictly arbitrary and always uncertain in their application. As far as the area-constant is concerned, we are accustomed to look upon it as accidental. Is it certain that our imaginary astronomers would do the same? If they were able to compare two different solar systems, they would get the idea that this constant may assume several different values. But I supposed at the outset, as I was entitled to do, that their system would appear isolated, and that they would see no star which was foreign to their system. Under these conditions they could only detect a single constant, which would have an absolutely invariable, unique value. They would be led no doubt to look upon it as an essential constant.

One word in passing to forestall an objection. The inhabitants of this imaginary world could neither observe nor define the area-constant as we do, because absolute longitudes escape their notice; but that would not prevent them from being rapidly led to remark a certain constant which would be naturally introduced into their equations, and which would be nothing but what we call the area-constant. But then what would happen? If the area-constant is regarded as essential, as dependent upon a law of nature, then in order to calculate the distances of the planets at any given moment it would be sufficient to know the initial values of these distances and those of their first derivatives. From this new point of view, distances will be determined by differential equations of the second order. Would this completely satisfy the minds of these astronomers? I think not. In the first place, they would very soon see that in differentiating their equations so as to raise them to a higher order, these equations would become much more simple, and they would be especially struck by the difficulty which arises from symmetry. They would have to admit different laws, according as the aggregate of the planets presented the figure of a certain polyhedron or rather of a regular polyhedron, and these consequences can only be escaped by regarding the area-constant as accidental. I have taken this particularexample, because I have imagined astronomers who would not be in the least concerned with terrestrial mechanics and whose vision would be bounded by the solar system. But our conclusions apply in all cases. Our universe is more extended than theirs, since we have fixed stars; but it, too, is very limited, so we might reason on the whole of our universe just as these astronomers do on their solar system. We thus see that we should be definitively led to conclude that the equations which define distances are of an order higher than the second. Why should this alarm us — why do we find it perfectly natural that the sequence of phenomena depends on initial values of the first derivatives of these distances, while we hesitate to admit that they may depend on the initial values of the second derivatives? It can only be because of mental habits created in us by the constant study of the generalised principle of inertia and of its consequences. The values of the distances at any given moment depend upon their initial values, on that of their first derivatives, and something else. What is that *something else*? If we do not want it to be merely one of the second derivatives, we have only the choice of hypotheses. Suppose, as is usually done, that this something else is the absolute orientation of the universe in space, or the rapidity with which this orientation varies; this may be, it certainly is, the most convenient solution for the geometer. But it is not the most satisfactory for the philosopher, because this orientation does not exist. We may assume that this something else is the position or the velocity of some invisible body, and this is what is done by certain persons, who have even called the body Alpha, although we are destined to never know anything about this body except its name. This is an artifice entirely analogous to that of which I spoke at the end of the paragraph containing my reflections on the principle of inertia. But as a matter of fact the difficulty is artificial. Provided that the future indications of our instruments can only depend on the indications which they have given us, or that they might have formerly given us, such is all we want, and with these conditions we may rest satisfied.

CHAPTER VIII.
ENERGY AND THERMO-DYNAMICS.

Energetics. — The difficulties raised by the classical mechanics have led certain minds to prefer a new system which they call Energetics. Energetics took its rise in consequence of the discovery of the principle of the conservation of energy. Helmholtz gave it its definite form. We begin by defining two quantities which play a fundamental part in this theory. They are *kinetic energy*, or *vis viva*, and *potential energy*. Every change that the bodies of nature can undergo is regulated by two experimental laws. First, the sum of the kinetic and potential energies is constant. This is the principle of the conservation of energy. Second, if a system of bodies is at A at the time t_0, and at B at the time t_1, it always passes from the first position to the second by such a path that the *mean* value of the difference between the two kinds of energy in the interval of time which separates the two epochs t_0 and t_1 is a minimum. This is Hamilton's principle, and is one of the forms of the principle of least action. The energetic theory has the following advantages over the classical. First, it is less incomplete — that is to say, the principles of the conservation of energy and of Hamilton teach us more than the fundamental principles of the classical theory, and exclude certain motions which do not occur in nature and which would be compatible with the classical theory. Second, it frees us from the hypothesis of atoms, which it was almost impossible to avoid with the classical theory. But in its turn it raises fresh difficulties. The definitions of the two kinds of energy would raise difficulties almost as great as those of force and mass in the first system. However, we can get out of these difficulties more easily, at any rate in the simplest cases. Assume an isolated system formed of a certain number of material points. Assume that these points are acted upon by forces depending only on their relative position and their distances apart, and independent of their velocities. In virtue of the principle of the conservation of energy there must be a function of forces. In this simple case the enunciation of the principle of the conservation of energy is of extreme simplicity. A certain quantity, which may be determined by experiment, must remain constant. This quantity is the sum of two terms. The first depends only on the position of the material points, and is independent of their velocities; the second is proportional to the squares of these velocities. This decomposition can only take place in one way. The first of these terms, which I shall call U, will be potential energy; the second, which I shall call T, will be kinetic energy. It is true that if T + U is constant, so is any function of T + U, φ (T + U). But this function φ (T + U) will not be the sum of two terms, the one independent of the velocities, and the other proportional to the square of the velocities. Among the functions which remain constant there is only one which enjoys this property. It is T + U (or a linear function of T + U), it matters not which, since this linear function may always be reduced to T + U by a change of unit and of origin. This, then, is what we call energy. The first term we shall call potential energy, and the second kinetic energy. The definition of the two kinds of energy may therefore be carried through without any ambiguity.

So it is with the definition of mass. Kinetic energy, or *vis viva*, is expressed very simply by the aid of the masses, and of the relative velocities of all the material points with reference to one of them. These relative velocities may be observed, and when we have the expression of the kinetic energy as a function of these relative velocities, the coefficients of this expression will give us the masses. So in this simple case the fundamental ideas can be defined without difficulty. But the difficulties reappear in the more complicated cases if the forces, instead of depending solely on the distances, depend also on the velocities. For example, Weber supposes the mutual action of two electric molecules to depend not only on their distance but on their velocity and on their acceleration. If material points attracted each other according to an analogous law, U would depend on the velocity, and it might contain a term proportional to the square of the velocity. How can we detect among such terms those that arise from T or U? and how, therefore, can we distinguish the two parts of the energy? But there is more than this. How can we define energy itself? We have no more reason to take as our definition T + U rather than any other function of T + U, when the property which characterised T + U has disappeared — namely, that of being the sum of two terms of a particular form. But that is not all. We must take account, not only of mechanical energy properly so called, but of the other forms of energy — heat, chemical energy, electrical energy, etc. The principle of the conservation of energy must be written T + U + Q = a constant, where T is the sensible kinetic energy, U the potential energy of position, depending only on the position of the bodies, Q the internal molecular energy under the thermal, chemical, or electrical form. This would be all right if the three terms were absolutely distinct; if T were proportional to the square of the velocities, U independent of these velocities and of the state of the bodies, Q independent of the velocities and of the positions of the bodies, and depending only on their internal state. The expression for the energy could be decomposed in one way only into three terms of this form. But this is not the case. Let us consider electrified bodies. The electro-static energy due to their mutual action will evidently depend on their charge — *i.e.*, on their state; but it will equally depend on their position. If these bodies are in motion, they will act electro-dynamically on one another, and the electro-dynamic energy will depend not only on their state and their position but on their velocities. We have therefore no means of making the selection of the terms which should form part of T, and U, and Q, and of separating the three parts of the energy. If T + U + Q is constant, the same is true of any function whatever, φ (T + U + Q).

If T + U + Q were of the particular form that I have suggested above, no ambiguity would ensue. Among the functions φ (T + U + Q) which remain constant, there is only one that would be of this particular form, namely the one which I would agree to call energy. But I have said this is not rigorously the case. Among the functions that remain constant there is not one which can rigorously be placed in this particular form. How then can we choose from among them that which should be called energy? We have no longer any guide in our choice.

Of the principle of the conservation of energy there is nothing left then but an enunciation: — *There is something which remains constant.* In this form it, in its turn, is outside the bounds of experiment and reduced to a kind of tautology. It is clear that if the world is governed by laws there will be quantities which remain constant. Like Newton's laws, and for an analogous reason, the principle of the conservation of energy being based on experiment, can no longer be invalidated by it.

This discussion shows that, in passing from the classical system to the energetic, an advance has been made; but it shows, at the same time, that we have not advanced far enough.

Another objection seems to be still more serious. The principle of least action is applicable to reversible phenomena, but it is by no means satisfactory as far as irreversible phenomena are concerned. Helmholtz attempted to extend it to this class of phenomena, but he did not and could not succeed. So far as this is concerned all has yet to be done. The very enunciation of the principle of least action is objectionable. To move from one point to another, a material molecule, acted upon by no force, but compelled to move on a surface, will take as its path the geodesic line — *i.e.*, the shortest path. This molecule seems to know the point to which we want to take it, to foresee the time that it will take it to reach it by such a path, and then to know how to choose the most convenient path. The enunciation of the principle presents it to us, so to speak, as a living and free entity. It is clear that it would be better to replace it by a less objectionable enunciation, one in which, as philosophers would say, final effects do not seem to be substituted for acting causes.

Thermo-dynamics. — The rôle of the two fundamental principles of thermo-dynamics becomes daily more important in all branches of natural philosophy. Abandoning the ambitious theories of forty years ago, encumbered as they were with molecular hypotheses, we now try to rest on thermo-dynamics alone the entire edifice of mathematical physics. Will the two principles of Mayer and of Clausius assure to it foundations solid enough to last for some time? We all feel it, but whence does our confidence arise? An eminent physicist said to me one day, *àpropos* of the law of errors: — every one stoutly believes it, because mathematicians imagine that it is an effect of observation, and observers imagine that it is a mathematical theorem. And this was for a long time the case with the principle of the conservation of energy. It is no longer the same now. There is no one who does not know that it is an experimental fact. But then who gives us the right of attributing to the principle itself more generality and more precision than to the experiments which have served to demonstrate it? This is asking, if it is legitimate to generalise, as we do every day, empiric data, and I shall not be so foolhardy as to discuss this question, after so many philosophers have vainly tried to solve it. One thing alone is certain. If this permission were refused to us, science could not exist; or at least would be reduced to a kind of inventory, to the ascertaining of isolated facts. It would not longer be to us of any value, since it could not satisfy our need of order and harmony, and because it would be at the same time incapable of prediction. As the circumstances which have preceded any fact whatever will never again, in all probability, be simultaneously reproduced, we already require a first generalisation to predict whether the fact will be renewed as soon as the least of these circumstances is changed. But every proposition may be generalised in an infinite number of ways. Among all possible generalisations we must choose, and we cannot but choose the simplest. We are therefore led to adopt the same course as if a simple law were, other things being equal, more probable than a complex law. A century ago it was frankly confessed and proclaimed abroad that Nature loves simplicity; but Nature has proved the contrary since then on more than one occasion. We no longer confess this tendency, and we only keep of it what is indispensable, so that science may not become impossible. In formulating a general, simple, and formal law, based on a comparatively small number of not altogether consistent experiments, we have only obeyed a necessity from which the human mind cannot free itself. But there is something more, and that is why I dwell on this topic. No one doubts that Mayer's principle is not called upon to survive all the particular laws from which it was deduced, in the same way that Newton's law has survived the laws of Kepler from which it was derived, and which are no longer anything but approximations, if we take perturbations into account. Now why does this principle thus occupy a kind of privileged position among physical laws? There are many reasons for that. At the outset we think that we cannot reject it, or even doubt its absolute rigour, without admitting the possibility of perpetual motion; we certainly feel distrust at such a prospect, and we believe ourselves less rash in affirming it than in denying it. That perhaps is not quite accurate. The impossibility of perpetual motion only implies the conservation of energy for reversible phenomena. The imposing simplicity of Mayer's principle equally contributes to strengthen our faith. In a law immediately deduced from experiments, such as Mariotte's law, this simplicity would rather appear to us a reason for distrust; but here this is no longer the case. We take elements which at the first glance are unconnected; these arrange themselves in an unexpected order, and form a harmonious whole. We cannot believe that this unexpected harmony is a mere result of chance. Our conquest appears to be valuable to us in proportion to the efforts it has cost, and we feel the more certain of having snatched its true secret from Nature in proportion as Nature has appeared more jealous of our attempts to discover it. But these are only small reasons. Before we raise Mayer's law to the dignity of an absolute principle, a deeper discussion is necessary. But if we embark on this discussion we see that this absolute principle is not even easy to enunciate. In every particular case we clearly see what energy is, and we can give it at least a provisory definition; but it is impossible to find a general definition of it. If we wish to enunciate the principle in all its generality and apply it to the universe, we see it vanish, so to speak, and nothing is left but this — *there is something which remains constant*. But has this a meaning? In the determinist hypothesis the state of the universe is determined by an extremely large number n of parameters, which I shall call $x_1, x_2, x_3, \ldots x_n$. As soon as we know at a given moment the values of these n parameters, we also know their derivatives with respect to time, and we can therefore calculate the values of these same parameters at an anterior or ulterior moment. In other words, these n parameters specify n differential equations of the first order. These equations have n-1 integrals, and therefore there are n-1 functions of $x_1, x_2, x_3, \ldots x_n$, which remain constant. If we say then, *there is something which remains constant*, we are only enunciating a tautology. We would be even embarrassed to decide which among all our integrals is that which should retain the name of energy. Besides, it is not in this sense that Mayer's principle is understood when it is applied to a limited system. We admit, then, that p of our n parameters vary independently so that we have only n - p relations, generally linear, between our n parameters and their derivatives. Suppose, for the sake of simplicity, that the sum of the work done by the external forces is zero, as well as that of all the quantities of heat given off from the interior: what will then be the meaning of our principle? *There is a combination of these n - p relations, of which the first member is an exact differential*; and then this differential vanishing in virtue of our n - p relations, its integral is a constant, and it is this integral which we call energy. But how can it be that there are several parameters whose variations are independent? That can only take place in the case of external forces (although we have supposed, for the sake of simplicity, that the algebraical sum of all the work done by these forces has vanished). If, in fact, the system were completely isolated from all external action, the values of our n parameters at a given moment would suffice to determine the state of the system at any ulterior moment whatever, provided that we still clung to the determinist hypothesis. We should therefore fall back on the same difficulty as before. If the future state of the system is not entirely determined by its present state, it is because it further depends on the state of bodies external to the system. But then, is it likely that there exist among the parameters x which define the state of the system of equations independent of this state of the external bodies? and if in certain cases we think we can find them, is it not only because of our ignorance, and because the influence of these bodies is too weak for our experiment to be able to detect it? If the system is not regarded as completely isolated, it is probable that the rigorously exact expression of its internal energy will depend upon the state of the external bodies. Again, I have supposed above that the sum of all the external work is zero, and if we wish to be free from this rather artificial restriction the enunciation becomes still more difficult. To formulate Mayer's principle by giving it an absolute meaning, we must extend it to the whole universe, and then we find ourselves face to face with the very difficulty we have endeavoured to avoid. To sum up, and to use ordinary language, the law of the conservation of energy can have only one significance, because there is in it a property common to all possible properties; but in the determinist hypothesis there is only one possible, and then the law has no meaning. In the indeterminist hypothesis, on the other hand, it would have a meaning even if we wished to regard it in an absolute sense. It would appear as a limitation imposed on freedom.

But this word warns me that I am wandering from the subject, and that I am leaving the domain of mathematics and physics. I check myself, therefore, and I wish to retain only one impression of the whole of this discussion, and that is, that Mayer's law is a form subtle enough for us to be able to put into it almost anything we like. I do not mean by that that it corresponds to no objective reality, nor that it is reduced to mere tautology; since, in each particular case, and provided we do not wish to extend it to the absolute, it has a perfectly clear meaning. This subtlety is a reason for believing that it will last long; and as, on the other hand, it will only disappear to be blended in a higher harmony, we may work with confidence and utilise it, certain beforehand that our work will not be lost.

Almost everything that I have just said applies to the principle of Clausius. What distinguishes it is, that it is expressed by an inequality. It will be said perhaps that it is the same with all physical laws, since their precision is always limited by errors of observation. But they at least claim to be first approximations, and we hope to replace them little by little by more exact laws. If, on the other hand, the principle of Clausius reduces to an inequality, this is not caused by the imperfection of our means of observation, but by the very nature of the question.

General Conclusions on Part III. — The principles of mechanics are therefore presented to us under two different aspects. On the one hand, there are truths founded on experiment, and verified approximately as far as almost isolated systems are concerned; on the other hand, there are postulates applicable to the whole of the universe and regarded as rigorously true. If these postulates possess a generality and a certainty which falsify the experimental truths from which they were deduced, it is because they reduce in final analysis to a simple convention that we have a right to make, because we are certain beforehand that no experiment can contradict it. This convention, however, is not absolutely arbitrary; it is not the child of our caprice. We admit it because certain experiments have shown us that it will be convenient, and thus is explained how experiment has built up the principles of mechanics, and why, moreover, it cannot reverse them. Take a comparison with geometry. The fundamental propositions of geometry, for instance, Euclid's postulate, are only conventions, and it is quite as unreasonable to ask if they are true or false as to ask if the metric system is true or false. Only, these conventions are convenient, and there are certain experiments which prove it to us. At the first glance, the analogy is complete, the rôle of experiment seems the same. We shall therefore be tempted to say, either mechanics must be looked upon as experimental science and then it should be the same with geometry; or, on the contrary, geometry is a deductive science, and then we can say the same of mechanics. Such a conclusion would be illegitimate. The experiments which have led us to adopt as more convenient the fundamental conventions of geometry refer to bodies which have nothing in common with those that are studied by geometry. They refer to the properties of solid bodies and to the propagation of light in a straight line. These are mechanical, optical experiments. In no way can they be regarded as geometrical experiments. And even the probable reason why our geometry seems convenient to us is, that our bodies, our hands, and our limbs enjoy the properties of solid bodies. Our fundamental experiments are pre-eminently physiological experiments which refer, not to the space which is the object that geometry must study, but to our body — that is to say, to the instrument which we use for that study. On the other hand, the fundamental conventions of mechanics and the experiments which prove to us that they are convenient, certainly refer to the same objects or to analogous objects. Conventional and general principles are the natural and direct generalisations of experimental and particular principles. Let it not be said that I am thus tracing artificial frontiers between the sciences; that I am separating by a barrier geometry properly so called from the study of solid bodies. I might just as well raise a barrier between experimental mechanics and the conventional mechanics of general principles. Who does not see, in fact, that by separating these two sciences we mutilate both, and that what will remain of the conventional mechanics when it is isolated will be but very little, and can in no way be compared with that grand body of doctrine which is called geometry.

We now understand why the teaching of mechanics should remain experimental. Thus only can we be made to understand the genesis of the science, and that is indispensable for a complete knowledge of the science itself. Besides, if we study mechanics, it is in order to apply it; and we can only apply it if it remains objective. Now, as we have seen, when principles gain in generality and certainty they lose in objectivity. It is therefore especially with the objective side of principles that we must be early familiarised, and this can only be by passing from the particular to the general, instead of from the general to the particular.

Principles are conventions and definitions in disguise. They are, however, deduced from experimental laws, and these laws have, so to speak, been erected into principles to which our mind attributes an absolute value. Some philosophers have generalised far too much. They have thought that the principles were the whole of science, and therefore that the whole of science was conventional. This paradoxical doctrine, which is called Nominalism, cannot stand examination. How can a law become a principle? It expressed a relation between two real terms, A and B; but it was not rigorously true, it was only approximate. We introduce arbitrarily an intermediate term, C, more or less imaginary, and C is *by definition* that which has with A *exactly* the relation expressed by the law. So our law is decomposed into an absolute and rigorous principle which expresses the relation of A to C, and an approximate experimental and revisable law which expresses the relation of C to B. But it is clear that however far this decomposition may be carried, laws will always remain. We shall now enter into the domain of laws properly so called.

PART IV.

NATURE.

CHAPTER IX.
HYPOTHESES IN PHYSICS.

The Rôle of Experiment and Generalisation. — Experiment is the sole source of truth. It alone can teach us something new; it alone can give us certainty. These are two points that cannot be questioned. But then, if experiment is every thing, what place is left for mathematical physics? What can experimental physics do with such an auxiliary — an auxiliary, moreover, which seems useless, and even may be dangerous?

However, mathematical physics exists. It has rendered undeniable service, and that is a fact which has to be explained. It is not sufficient merely to observe; we must use our observations, and for that purpose we must generalise. This is what has always been done, only as the recollection of past errors has made man more and more circumspect, he has observed more and more and generalised less and less. Every age has scoffed at its predecessor, accusing it of having generalised too boldly and too naïvely. Descartes used to commiserate the Ionians. Descartes in his turn makes us smile, and no doubt some day our children will laugh at us. Is there no way of getting at once to the gist of the matter, and thereby escaping the raillery which we foresee? Cannot we be content with experiment alone? No, that is impossible; that would be a complete misunderstanding of the true character of science. The man of science must work with method. Science is built up of facts, as a house is built of stones; but an accumulation of facts is no more a science than a heap of stones is a house. Most important of all, the man of science must exhibit foresight. Carlyle has written somewhere something after this fashion. "Nothing but facts are of importance. John Lackland passed by here. Here is something that is admirable. Here is a reality for which I would give all the theories in the world."[1] Carlyle was a compatriot of Bacon, and, like him, he wished to proclaim his worship of *the God of Things as they are*.

But Bacon would not have said that. That is the language of the historian. The physicist would most likely have said: "John Lackland passed by here. It is all the same to me, for he will not pass this way again."

We all know that there are good and bad experiments. The latter accumulate in vain. Whether there are a hundred or a thousand, one single piece of work by a real master — by a Pasteur, for example — will be sufficient to sweep them into oblivion. Bacon would have thoroughly understood that, for he invented the phrase *experimentum crucis*; but Carlyle would not have under stood it. A fact is a fact. A student has read such and such a number on his thermometer. He has taken no precautions. It does not matter; he has read it, and if it is only the fact which counts, this is a reality that is as much entitled to be called a reality as the peregrinations of King John Lackland. What, then, is a good experiment? It is that which teaches us something more than an isolated fact. It is that which enables us to predict, and to generalise. Without generalisation, prediction is impossible. The circumstances under which one has operated will never again be reproduced simultaneously. The fact observed will never be repeated. All that can be affirmed is that under analogous circumstances an analogous fact will be produced. To predict it, we must therefore invoke the aid of analogy — that is to say, even at this stage, we must generalise. However timid we may be, there must be interpolation. Experiment only gives us a certain number of isolated points. They must be connected by a continuous line, and this is a true generalisation. But more is done. The curve thus traced will pass between and near the points observed; it will not pass through the points themselves. Thus we are not restricted to generalising our experiment, we correct it; and the physicist who would abstain from these corrections, and really content himself with experiment pure and simple, would be compelled to enunciate very extraordinary laws indeed. Detached facts cannot therefore satisfy us, and that is why our science must be ordered, or, better still, generalised.

It is often said that experiments should be made without preconceived ideas. That is impossible. Not only would it make every experiment fruitless, but even if we wished to do so, it could not be done. Every man has his own conception of the world, and this he cannot so easily lay aside. We must, for example, use language, and our language is necessarily steeped in preconceived ideas. Only they are unconscious preconceived ideas, which are a thousand times the most dangerous of all. Shall we say, that if we cause others to intervene of which we are fully conscious, that we shall only aggravate the evil? I do not think so. I am inclined to think that they will serve as ample counterpoises — I was almost going to say antidotes. They will generally disagree, they will enter into conflict one with another, and *ipso facto*, they will force us to look at things under different aspects. This is enough to free us. He is no longer a slave who can choose his master.

Thus, by generalisation, every fact observed enables us to predict a large number of others; only, we ought not to forget that the first alone is certain, and that all the others are merely probable. However solidly founded a prediction may appear to us, we are never *absolutely* sure that experiment will not prove it to be baseless if we set to work to verify it. But the probability of its accuracy is often so great that practically we may be content with it. It is far better to predict without certainty, than never to have predicted at all. We should never, therefore, disdain to verify when the opportunity presents itself. But every experiment is long and difficult, and the labourers are few, and the number of facts which we require to predict is enormous; and besides this mass, the number of direct verifications that we can make will never be more than a negligible quantity. Of this little that we can directly attain we must choose the best. Every experiment must enable us to make a maximum number of predictions having the highest possible degree of probability. The problem is, so to speak, to increase the output of the scientific machine. I may be permitted to compare science to a library which must go on increasing indefinitely; the librarian has limited funds for his purchases, and he must, therefore, strain every nerve not to waste them. Experimental physics has to make the purchases, and experimental physics alone can enrich the library. As for mathematical physics, her duty is to draw up the catalogue. If the catalogue is well done the library is none the richer for it; but the reader will be enabled to utilise its riches; and also by showing the librarian the gaps in his collection, it will help him to make a judicious use of his funds, which is all the more important, inasmuch as those funds are entirely inadequate. That is the rôle of mathematical physics. It must direct generalisation, so as to increase what I called just now the output of science. By what means it does this, and how it may do it without danger, is what we have now to examine.

The Unity of Nature. — Let us first of all observe that every generalisation supposes in a certain measure a belief in the unity and simplicity of Nature. As far as the unity is concerned, there can be no difficulty. If the different parts of the universe were not as the organs of the same body, they would not re-act one upon the other; they would mutually ignore each other, and we in particular should only know one part. We need not, therefore, ask if Nature is one, but how she is one.

As for the second point, that is not so clear. It is not certain that Nature is simple. Can we without danger act as if she were?

There was a time when the simplicity of Mariotte's law was an argument in favour of its accuracy: when Fresnel himself, after having said in a conversation with Laplace that Nature cares naught for analytical difficulties, was compelled to explain his words so as not to give offence to current opinion. Nowadays, ideas have changed considerably; but those who do not believe that natural laws must be simple, are still often obliged to act as if they did believe it. They cannot entirely dispense with this necessity without making all generalisation, and therefore all science, impossible. It is clear that any fact can be generalised in an infinite number of ways, and it is a question of choice. The choice can only be guided by considerations of simplicity. Let us take the most ordinary case, that of interpolation. We draw a continuous line as regularly as possible between the points given by observation. Why do we avoid angular points and inflexions that are too sharp? Why do we not make our curve describe the most capricious zigzags? It is because we know beforehand, or think we know, that the law we have to express cannot be so complicated as all that. The mass of Jupiter may be deduced either from the movements of his satellites, or from the perturbations of the major planets, or from those of the minor planets. If we take the mean of the determinations obtained by these three methods, we find three numbers very close together, but not quite identical. This result might be interpreted by supposing that the gravitation constant is not the same in the three cases; the observations would be certainly much better represented. Why do we reject this interpretation? Not because it is absurd, but because it is uselessly complicated. We shall only accept it when we are forced to, and it is not imposed upon us yet. To sum up, in most cases every law is held to be simple until the contrary is proved.

This custom is imposed upon physicists by the reasons that I have indicated, but how can it be justified in the presence of discoveries which daily show us fresh details, richer and more complex? How can we even reconcile it with the unity of nature? For if all things are interdependent, the relations in which so many different objects intervene can no longer be simple.

If we study the history of science we see produced two phenomena which are, so to speak, each the inverse of the other. Sometimes it is simplicity which is hidden under what is apparently complex; sometimes, on the contrary, it is simplicity which is apparent, and which conceals extremely complex realities. What is there more complicated than the disturbed motions of the planets, and what more simple than Newton's law? There, as Fresnel said, Nature playing with analytical difficulties, only uses simple means, and creates by their combination I know not what tangled skein. Here it is the hidden simplicity which must be disentangled. Examples to the contrary abound. In the kinetic theory of gases, molecules of tremendous velocity are discussed, whose paths, deformed by incessant impacts, have the most capricious shapes, and plough their way through space in every direction. The result observable is Mariotte's simple law. Each individual fact was complicated. The law of great numbers has re-established simplicity in the mean. Here the simplicity is only apparent, and the coarseness of our senses alone prevents us from seeing the complexity.

Many phenomena obey a law of proportionality. But why? Because in these phenomena there is something which is very small. The simple law observed is only the translation of the general analytical rule by which the infinitely small increment of a function is proportional to the increment of the variable. As in reality our increments are not infinitely small, but only very small, the law of proportionality is only approximate, and simplicity is only apparent. What I have just said applies to the law of the superposition of small movements, which is so fruitful in its applications and which is the foundation of optics.

And Newton's law itself? Its simplicity, so long undetected, is perhaps only apparent. Who knows if it be not due to some complicated mechanism, to the impact of some subtle matter animated by irregular movements, and if it has not become simple merely through the play of averages and large numbers? In any case, it is difficult not to suppose that the true law contains complementary terms which may become sensible at small distances. If in astronomy they are negligible, and if the law thus regains its simplicity, it is solely on account of the enormous distances of the celestial bodies. No doubt, if our means of investigation became more and more penetrating, we should discover the simple beneath the complex, and then the complex from the simple, and then again the simple beneath the complex, and so on, without ever being able to predict what the last term will be. We must stop somewhere, and for science to be possible we must stop where we have found simplicity. That is the only ground on which we can erect the edifice of our generalisations. But, this simplicity being only apparent, will the ground be solid enough? That is what we have now to discover.

For this purpose let us see what part is played in our generalisations by the belief in simplicity. We have verified a simple law in a considerable number of particular cases. We refuse to admit that this coincidence, so often repeated, is a result of mere chance, and we conclude that the law must be true in the general case.

Kepler remarks that the positions of a planet observed by Tycho are all on the same ellipse. Not for one moment does he think that, by a singular freak of chance, Tycho had never looked at the heavens except at the very moment when the path of the planet happened to cut that ellipse. What does it matter then if the simplicity be real or if it hide a complex truth? Whether it be due to the influence of great numbers which reduces individual differences to a level, or to the greatness or the smallness of certain quantities which allow of certain terms to be neglected — in no case is it due to chance. This simplicity, real or apparent, has always a cause. We shall therefore always be able to reason in the same fashion, and if a simple law has been observed in several particular cases, we may legitimately suppose that it still will be true in analogous cases. To refuse to admit this would be to attribute an inadmissible rôle to chance. However, there is a difference. If the simplicity were real and profound it would bear the test of the increasing precision of our methods of measurement. If, then, we believe Nature to be profoundly simple, we must conclude that it is an approximate and not a rigorous simplicity. This is what was formerly done, but it is what we have no longer the right to do. The simplicity of Kepler's laws, for instance, is only apparent; but that does not prevent them from being applied to almost all systems analogous to the solar system, though that prevents them from being rigorously exact.

Rôle of Hypothesis. — Every generalisation is a hypothesis. Hypothesis therefore plays a necessary rôle, which no one has ever contested. Only, it should always be as soon as possible submitted to verification. It goes without saying that, if it cannot stand this test, it must be abandoned without any hesitation. This is, indeed, what is generally done; but sometimes with a certain impatience. Ah well! this impatience is not justified. The physicist who has just given up one of his hypotheses should, on the contrary, rejoice, for he found an unexpected opportunity of discovery. His hypothesis, I imagine, had not been lightly adopted. It took into account all the known factors which seem capable of intervention in the phenomenon. If it is not verified, it is because there is something unexpected and extraordinary about it, because we are on the point of finding something unknown and new. Has the hypothesis thus rejected been sterile? Far from it. It may be even said that it has rendered more service than a true hypothesis. Not only has it been the occasion of a decisive experiment, but if this experiment had been made by chance, without the hypothesis, no conclusion could have been drawn; nothing extraordinary would have been seen; and only one fact the more would have been catalogued, without deducing from it the remotest consequence.

Now, under what conditions is the use of hypothesis without danger? The proposal to submit all to experiment is not sufficient. Some hypotheses are dangerous, — first and foremost those which are tacit and unconscious. And since we make them without knowing them, we cannot get rid of them. Here again, there is a service that mathematical physics may render us. By the precision which is its characteristic, we are compelled to formulate all the hypotheses that we would unhesitatingly make without its aid. Let us also notice that it is important not to multiply hypotheses indefinitely. If we construct a theory based upon multiple hypotheses, and if experiment condemns it, which of the premises must be changed? It is impossible to tell. Conversely, if the experiment succeeds, must we suppose that it has verified all these hypotheses at once? Can several unknowns be determined from a single equation?

We must also take care to distinguish between the different kinds of hypotheses. First of all, there are those which are quite natural and necessary. It is difficult not to suppose that the influence of very distant bodies is quite negligible, that small movements obey a linear law, and that effect is a continuous function of its cause. I will say as much for the conditions imposed by symmetry. All these hypotheses affirm, so to speak, the common basis of all the theories of mathematical physics. They are the last that should be abandoned. There is a second category of hypotheses which I shall qualify as indifferent. In most questions the analyst assumes, at the beginning of his calculations, either that matter is continuous, or the reverse, that it is formed of atoms. In either case, his results would have been the same. On the atomic supposition he has a little more difficulty in obtaining them — that is all. If, then, experiment confirms his conclusions, will he suppose that he has proved, for example, the real existence of atoms?

In optical theories two vectors are introduced, one of which we consider as a velocity and the other as a vortex. This again is an indifferent hypothesis, since we should have arrived at the same conclusions by assuming the former to be a vortex and the latter to be a velocity. The success of the experiment cannot prove, therefore, that the first vector is really a velocity. It only proves one thing — namely, that it is a vector; and that is the only hypothesis that has really been introduced into the premises. To give it the concrete appearance that the fallibility of our minds demands, it was necessary to consider it either as a velocity or as a vortex. In the same way, it was necessary to represent it by an x or a y, but the result will not prove that we were right or wrong in regarding it as a velocity; nor will it prove we are right or wrong in calling it x and not y.

These indifferent hypotheses are never dangerous provided their characters are not misunderstood. They may be useful, either as artifices for calculation, or to assist our understanding by concrete images, to fix the ideas, as we say. They need not therefore be rejected. The hypotheses of the third category are real generalisations. They must be confirmed or invalidated by experiment. Whether verified or condemned, they will always be fruitful; but, for the reasons I have given, they will only be so if they are not too numerous.

Origin of Mathematical Physics. — Let us go further and study more closely the conditions which have assisted the development of mathematical physics. We recognise at the outset that the efforts of men of science have always tended to resolve the complex phenomenon given directly by experiment into a very large number of elementary phenomena, and that in three different ways.

First, with respect to time. Instead of embracing in its entirety the progressive development of a phenomenon, we simply try to connect each moment with the one immediately preceding. We admit that the present state of the world only depends on the immediate past, without being directly influenced, so to speak, by the recollection of a more distant past. Thanks to this postulate, instead of studying directly the whole succession of phenomena, we may confine ourselves to writing down its *differential equation*; for the laws of Kepler we substitute the law of Newton.

Next, we try to decompose the phenomena in space. What experiment gives us is a confused aggregate of facts spread over a scene of consider able extent. We must try to deduce the elementary phenomenon, which will still be localised in a very small region of space.

A few examples perhaps will make my meaning clearer. If we wished to study in all its complexity the distribution of temperature in a cooling solid, we could never do so. This is simply be cause, if we only reflect that a point in the solid can directly impart some of its heat to a neighbouring point, it will immediately impart that heat only to the nearest points, and it is but gradually that the flow of heat will reach other portions of the solid. The elementary phenomenon is the interchange of heat between two contiguous points. It is strictly localised and relatively simple if, as is natural, we admit that it is not influenced by the temperature of the molecules whose distance apart is small.

I bend a rod: it takes a very complicated form, the direct investigation of which would be impossible. But I can attack the problem, however, if I notice that its flexure is only the resultant of the deformations of the very small elements of the rod, and that the deformation of each of these elements only depends on the forces which are directly applied to it, and not in the least on those which may be acting on the other elements.

In all these examples, which may be increased without difficulty, it is admitted that there is no action at a distance or at great distances. That is an hypothesis. It is not always true, as the law of gravitation proves. It must therefore be verified. If it is confirmed, even approximately, it is valuable, for it helps us to use mathematical physics, at any rate by successive approximations. If it does not stand the test, we must seek something else that is analogous, for there are other means of arriving at the elementary phenomenon. If several bodies act simultaneously, it may happen that their actions are independent, and may be added one to the other, either as vectors or as scalar quantities. The elementary phenomenon is then the action of an isolated body. Or suppose, again, it is a question of small movements, or more generally of small variations which obey the well- known law of mutual or relative independence. The movement observed will then be decomposed into simple movements — for example, sound into its harmonics, and white light into its monochromatic components. When we have discovered in which direction to seek for the elementary phenomena, by what means may we reach it? First, it will often happen that in order to predict it, or rather in order to predict what is useful to us, it will not be necessary to know its mechanism. The law of great numbers will suffice. Take for example the propagation of heat. Each molecule radiates towards its neighbour — we need not inquire according to what law; and if we make any supposition in this respect, it will be an indifferent hypothesis, and therefore useless and unverifiable. In fact, by the action of averages and thanks to the symmetry of the medium, all differences are levelled, and, whatever the hypothesis may be, the result is always the same.

The same feature is presented in the theory of elasticity, and in that of capillarity. The neighbouring molecules attract and repel each other, we need not inquire by what law. It is enough for us that this attraction is sensible at small distances only, and that the molecules are very numerous, that the medium is symmetrical, and we have only to let the law of great numbers come into play.

Here again the simplicity of the elementary phenomenon is hidden beneath the complexity of the observable resultant phenomenon; but in its turn this simplicity was only apparent and disguised a very complex mechanism. Evidently the best means of reaching the elementary phenomenon would be experiment. It would be necessary by experimental artifices to dissociate the complex system which nature offers for our investigations and carefully to study the elements as dissociated as possible; for example, natural white light would be decomposed into monochromatic lights by the aid of the prism, and into polarised lights by the aid of the polariser. Unfortunately, that is neither always possible nor always sufficient, and sometimes the mind must run ahead of experiment. I shall only give one example which has always struck me rather forcibly. If I de compose white light, I shall be able to isolate a portion of the spectrum, but however small it may be, it will always be a certain width. In the same way the natural lights which are called *monochromatic* give us a very fine array, but a y which is not, however, infinitely fine. It might be supposed that in the experimental study of the properties of these natural lights, by operating with finer and finer rays, and passing on at last to the limit, so to speak, we should eventually obtain the properties of a rigorously monochromatic light. That would not be accurate. I assume that two rays emanate from the same source, that they are first polarised in planes at right angles, that they are then brought back again to the same plane of polarisation, and that we try to obtain interference. If the light were rigorously monochromatic, there would be interference; but with our nearly monochromatic lights, there will be no interference, and that, however narrow the ray may be. For it to be otherwise, the ray would have to be several million times finer than the finest known rays.

Here then we should be led astray by proceeding to the limit. The mind has to run ahead of the experiment, and if it has done so with success, it is because it has allowed itself to be guided by the instinct of simplicity. The knowledge of the elementary fact enables us to state the problem in the form of an equation. It only remains to deduce from it by combination the observable and verifiable complex fact. That is what we call *integration*, and it is the province of the mathematician. It might be asked, why in physical science generalisation so readily takes the mathematical form. The reason is now easy to see. It is not only because we have to express numerical laws; it is because the observable phenomenon is due to the superposition of a large number of elementary phenomena which are *all similar to each other*; and in this way differential equations are quite naturally introduced. It is not enough that each elementary phenomenon should obey simple laws: all those that we have to combine must obey the same law; then only is the intervention of mathematics of any use. Mathematics teaches us, in fact, to combine like with like. Its object is to divine the result of a combination without having to reconstruct that combination element by element. If we have to repeat the same operation several times, mathematics enables us to avoid this repetition by telling the result beforehand by a kind of induction. This I have explained before in the chapter on mathematical reasoning. But for that purpose all these operations must be similar; in the contrary case we must evidently make up our minds to working them out in full one after the other, and mathematics will be useless. It is therefore, thanks to the approximate homogeneity of the matter studied by physicists, that mathematical physics came into existence. In the natural sciences the following conditions are no longer to be found: — homogeneity, relative independence of remote parts, simplicity of the elementary fact; and that is why the student of natural science is compelled to have recourse to other modes   of   generalisation.

Footnotes

1.   V. *Past and Present*, end of Chapter I., Book II. — [TR.]

CHAPTER X.
THE THEORIES OF MODERN PHYSICS.

Significance of Physical Theories. — The ephemeral nature of scientific theories takes by surprise the man of the world. Their brief period of prosperity ended, he sees them abandoned one after another; he sees ruins piled upon ruins; he predicts that the theories in fashion to-day will in a short time succumb in their turn, and he concludes that they are absolutely in vain. This is what he calls the *bankruptcy of science.*

His scepticism is superficial; he does not take into account the object of scientific theories and the part they play, or he would understand that the ruins may be still good for something. No theory seemed established on firmer ground than Fresnel's, which attributed light to the movements of the ether. Then if Maxwell's theory is to-day preferred, does that mean that Fresnel's work was in vain? No; for Fresnel's object was not to know whether there really is an ether, if it is or is not formed of atoms, if these atoms really move in this way or that; his object was to predict optical phenomena.

This Fresnel's theory enables us to do to-day as well as it did before Maxwell's time. The differential equations are always true, they may be always integrated by the same methods, and the results of this integration still preserve their value. It cannot be said that this is reducing physical theories to simple practical recipes; these equations express relations, and if the equations remain true, it is because the relations preserve their reality. They teach us now, as they did then, that there is such and such a relation between this thing and that; only, the something which we then called *motion*, we now call electric *current*. But these are merely names of the images we substituted for the real objects which Nature will hide for ever from our eyes. The true relations between these real objects are the only reality we can attain, and the sole condition is that the same relations shall exist between these objects as between the images we are forced to put in their place. If the relations are known to us, what does it matter if we think it convenient to replace one image by another?

That a given periodic phenomenon (an electric oscillation, for instance) is really due to the vibration of a given atom, which, behaving like a pendulum, is really displaced in this manner or that, all this is neither certain nor essential. But that there is between the electric oscillation, the movement of the pendulum, and all periodic phenomena an intimate relationship which corresponds to a profound reality; that this relationship, this similarity, or rather this parallelism, is continued in the details; that it is a consequence of more general principles such as that of the conservation of energy, and that of least action; this we may affirm; this is the truth which will ever remain the same in whatever garb we may see fit to clothe it.

Many theories of dispersion have been proposed. The first were imperfect, and contained but little truth. Then came that of Helmholtz, and this in its turn was modified in different ways; its author himself conceived another theory, founded on Maxwell's principles. But the remarkable thing is, that all the scientists who followed Helmholtz obtain the same equations, although their starting-points were to all appearance widely separated. I venture to say that these theories are all simultaneously true; not merely because they express a true relation — that between absorption and abnormal dispersion. In the premises of these theories the part that is true is the part common to all: it is the affirmation of this or that relation between certain things, which some call by one name and some by another.

The kinetic theory of gases has given rise to many objections, to which it would be difficult to find an answer were it claimed that the theory is absolutely true. But all these objections do not alter the fact that it has been useful, particularly in revealing to us one true relation which would otherwise have remained profoundly hidden — the relation between gaseous and osmotic pressures. In this sense, then, it may be said to be true.

When a physicist finds a contradiction between two theories which are equally dear to him, he sometimes says: "Let us not be troubled, but let us hold fast to the two ends of the chain, lest we lose the intermediate links." This argument of the embarrassed theologian would be ridiculous if we were to attribute to physical theories the interpretation given them by the man of the world. In case of contradiction one of them at least should be considered false. But this is no longer the case if we only seek in them what should be sought. It is quite possible that they both express true relations, and that the contradictions only exist in the images we have formed to ourselves of reality. To those who feel that we are going too far in our limitations of the domain accessible to the scientist, I reply: These questions which we forbid you to investigate, and which you so regret, are not only insoluble, they are illusory and devoid of meaning.

Such a philosopher claims that all physics can be explained by the mutual impact of atoms. If he simply means that the same relations obtain between physical phenomena as between the mutual impact of a large number of billiard balls — well and good! this is verifiable, and perhaps is true. But he means something more, and we think we understand him, because we think we know what an impact is. Why? Simply because we have often watched a game of billiards. Are we to understand that God experiences the same sensations in the contemplation of His work that we do in watching a game of billiards? If it is not our intention to give his assertion this fantastic meaning, and if we do not wish to give it the more restricted meaning I have already mentioned, which is the sound meaning, then it has no meaning at all. Hypotheses of this kind have therefore only a metaphorical sense. The scientist should no more banish them than a poet banishes metaphor; but he ought to know what they are worth. They may be useful to give satisfaction to the mind, and they will do no harm as long as they are only indifferent hypotheses.

These considerations explain to us why certain theories, that were thought to be abandoned and definitively condemned by experiment, are suddenly revived from their ashes and begin a new life. It is because they expressed true relations, and had not ceased to do so when for some reason or other we felt it necessary to enunciate the same relations in another language. Their life had been latent, as it were.

Barely fifteen years ago, was there anything more ridiculous, more quaintly old-fashioned, than the fluids of Coulomb? And yet, here they are re-appearing under the name of *electrons*. In what do these permanently electrified molecules differ from the electric molecules of Coulomb? It is true that in the electrons the electricity is supported by a little, a very little matter; in other words, they have mass. Yet Coulomb did not deny mass to his fluids, or if he did, it was with reluctance. It would be rash to affirm that the belief in electrons will not also undergo an eclipse, but it was none the less curious to note this unexpected renaissance.

But the most striking example is Carnot's principle. Carnot established it, starting from false hypotheses. When it was found that heat was indestructible, and may be converted into work, his ideas were completely abandoned; later, Clausius returned to them, and to him is due their definitive triumph. In its primitive form, Carnot's theory expressed in addition to true relations, other inexact relations, the *débris* of old ideas; but the presence of the latter did not alter the reality of the others.

Clausius had only to separate them, just as one lops off dead branches.

The result was the second fundamental law of thermodynamics. The relations were always the same, although they did not hold, at least to all appearance, between the same objects. This was sufficient for the principle to retain its value. Nor have the reasonings of Carnot perished on this account; they were applied to an imperfect conception of matter, but their form — *i.e.*, the essential part of them, remained correct. What I have just said throws some light at the same time on the rôle of general principles, such as those of the principle of least action or of the conservation of energy. These principles are of very great value. They were obtained in the search for what there was in common in the enunciation of numerous physical laws; they thus represent the quintessence of innumerable observations. However, from their very generality results a consequence to which I have called attention in Chapter VIII. — namely, that they are no longer capable of verification. As we cannot give a general definition of energy, the principle of the conservation of energy simply signifies that there is a *something* which remains constant. Whatever fresh notions of the world may be given us by future experiments, we are certain beforehand that there is something which remains constant, and which may be called *energy*. Does this mean that the principle has no meaning and vanishes into a tautology? Not at all. It means that the different things to which we give the name of *energy* are connected by a true relationship; it affirms between them a real relation. But then, if this principle has a meaning, it may be false; it may be that we have no right to extend indefinitely its applications, and yet it is certain beforehand to be verified in the strict sense of the word. How, then, shall we know when it has been extended as far as is legitimate? Simply when it ceases to be useful to us — *i.e.*, when we can no longer use it to predict correctly new phenomena. We shall be certain in such a case that the relation affirmed is no longer real, for otherwise it would be fruitful; experiment without directly contradicting a new extension of the principle will nevertheless have condemned it.

Physics and Mechanism. — Most theorists have a constant predilection for explanations borrowed from physics, mechanics, or dynamics. Some would be satisfied if they could account for all phenomena by the motion of molecules attracting one another according to certain laws. Others are more exact: they would suppress attractions acting at a distance; their molecules would follow rectilinear paths, from which they would only be deviated by impacts. Others again, such as Hertz, suppress the forces as well, but suppose their molecules subjected to geometrical connections analogous, for instance, to those of articulated systems; thus, they wish to reduce dynamics to a kind of kinematics. In a word, they all wish to bend nature into a certain form, and unless they can do this they cannot be satisfied. Is Nature flexible enough for this?

We shall examine this question in Chapter XII., àpropos of Maxwell's theory. Every time that the principles of least action and energy are satisfied, we shall see that not only is there always a mechanical explanation possible, but that there is an unlimited number of such explanations. By means of a well-known theorem due to Königs, it may be shown that we can explain everything in an unlimited number of ways, by connections after the manner of Hertz, or, again, by central forces. No doubt it may be just as easily demonstrated that everything may be explained by simple impacts. For this, let us bear in mind that it is not enough to be content with the ordinary matter of which we are aware by means of our senses, and the movements of which we observe directly. We may conceive of ordinary matter as either composed of atoms, whose internal movements escape us, our senses being able to estimate only the displacement of the whole; or we may imagine one of those subtle fluids, which under the name of *ether* or other names, have from all time played so important a rôle in physical theories. Often we go further, and regard the ether as the only primitive, or even as the only true matter. The more moderate consider ordinary matter to be condensed ether, and there is nothing startling in this conception; but others only reduce its importance still further, and see in matter nothing more than the geometrical locus of singularities in the ether. Lord Kelvin, for instance, holds what we call matter to be only the locus of those points at which the ether is animated by vortex motions. Riemann believes it to be locus of those points at which ether is constantly destroyed; to Wiechert or Larmor, it is the locus of the points at which the ether has undergone a kind of torsion of a very particular kind. Taking any one of these points of view, I ask by what right do we apply to the ether the mechanical properties observed in ordinary matter, which is but false matter? The ancient fluids, caloric, electricity, etc., were abandoned when it was seen that heat is not indestructible. But they were also laid aside for another reason. In materialising them, their individuality was, so to speak, emphasised — gaps were opened between them; and these gaps had to be filled in when the sentiment of the unity of Nature became stronger, and when the intimate relations which connect all the parts were perceived. In multiplying the fluids, not only did the ancient physicists create unnecessary entities, but they destroyed real ties. It is not enough for a theory not to affirm false relations; it must not conceal true relations.

Does our ether actually exist? We know the origin of our belief in the ether. If light takes several years to reach us from a distant star, it is no longer on the star, nor is it on the earth. It must be somewhere, and supported, so to speak, by some material agency.

The same idea may be expressed in a more mathematical and more abstract form. What we note are the changes undergone by the material molecules. We see, for instance, that the photographic plate experiences the consequences of a phenomenon of which the incandescent mass of a star was the scene several years before. Now, in ordinary mechanics, the state of the system under consideration depends only on its state at the moment immediately preceding; the system therefore satisfies certain differential equations. On the other hand, if we did not believe in the ether, the state of the material universe would depend not only on the state immediately preceding, but also on much older states; the system would satisfy equations of finite differences. The ether was invented to escape this breaking down of the laws of general mechanics.

Still, this would only compel us to fill the interplanetary space with ether, but not to make it penetrate into the midst of the material media. Fizeau's experiment goes further. By the interference of rays which have passed through the air or water in motion, it seems to show us two different media penetrating each other, and yet being displaced with respect to each other. The ether is all but in our grasp. Experiments can be conceived in which we come closer still to it. Assume that Newton's principle of the equality of action and re-action is not true if applied to matter *alone*, and that this can be proved. The geometrical sum of all the forces applied to all the molecules would no longer be zero. If we did not wish to change the whole of the science of mechanics, we should have to introduce the ether, in order that the action which matter apparently undergoes should be counterbalanced by the re-action of matter on something. Or again, suppose we discover that optical and electrical phenomena are influenced by the motion of the earth. It would follow that those phenomena might reveal to us not only the relative motion of material bodies, but also what would seem to be their absolute motion. Again, it would be necessary to have an ether in order that these so-called absolute movements should not be their displacements with respect to empty space, but with respect to something concrete.

Will this ever be accomplished? I do not think so, and I shall explain why; and yet, it is not absurd, for others have entertained this view. For instance, if the theory of Lorentz, of which I shall speak in more detail in Chapter XIII., were true, Newton's principle would not apply to matter *alone*, and the difference would not be very far from being within reach of experiment. On the other hand, many experiments have been made on the influence of the motion of the earth. The results have always been negative. But if these experiments have been undertaken, it is because we have not been certain beforehand; and indeed, according to current theories, the compensation would be only approximate, and we might expect to find accurate methods giving positive results. I think that such a hope is illusory; it was none the less interesting to show that a success of this kind would, in a certain sense, open to us a new world.

And now allow me to make a digression; I must explain why I do not believe, in spite of Lorentz, that more exact observations will ever make evident anything else but the relative displacements of material bodies. Experiments have been made that should have disclosed the terms of the first order; the results were nugatory. Could that have been by chance? No one has admitted this; a general explanation was sought, and Lorentz found it. He showed that the terms of the first order should cancel each other, but not the terms of the second order. Then more exact experiments were made, which were also negative; neither could this be the result of chance. An explanation was necessary, and was forthcoming; they always are; hypotheses are what we lack the least. But this is not enough. Who is there who does not think that this leaves to chance far too important a role? Would it not also be a chance that this singular concurrence should cause a certain circumstance to destroy the terms of the first order, and that a totally different but very opportune circumstance should cause those of the second order to vanish? No; the same explanation must be found for the two cases, and everything tends to show that this explanation would serve equally well for the terms of the higher order, and that the mutual destruction of these terms will be rigorous and absolute.

The Present State of Physics. — Two opposite tendencies may be distinguished in the history of the development of physics. On the one hand, new relations are continually being discovered between objects which seemed destined to remain for ever unconnected; scattered facts cease to be strangers to each other and tend to be marshalled into an imposing synthesis. The march of science is towards unity and simplicity.

On the other hand, new phenomena are continually being revealed; it will be long before they can be assigned their place — sometimes it may happen that to find them a place a corner of the edifice must be demolished. In the same way, we are continually perceiving details ever more varied in the phenomena we know, where our crude senses used to be unable to detect any lack of unity. What we thought to be simple becomes complex, and the march of science seems to be towards diversity and complication.

Here, then, are two opposing tendencies, each of which seems to triumph in turn. Which will win? If the first wins, science is possible; but nothing proves this *à priori*, and it may be that after unsuccessful efforts to bend Nature to our ideal of unity in spite of herself, we shall be submerged by the ever-rising flood of our new riches and compelled to renounce all idea of classification — to abandon our ideal, and to reduce science to the mere recording of innumerable recipes.

In fact, we can give this question no answer. All that we can do is to observe the science of to-day, and compare it with that of yesterday. No doubt after this examination we shall be in a position to offer a few conjectures.

Half-a-century ago hopes ran high indeed. The unity of force had just been revealed to us by the discovery of the conservation of energy and of its transformation. This discovery also showed that the phenomena of heat could be explained by molecular movements. Although the nature of these movements was not exactly known, no one doubted but that they would be ascertained before long. As for light, the work seemed entirely completed. So far as electricity was concerned, there was not so great an advance. Electricity had just annexed magnetism. This was a considerable and a definitive step towards unity. But how was electricity in its turn to be brought into the general unity, and how was it to be included in the general universal mechanism? No one had the slightest idea. As to the possibility of the inclusion, all were agreed; they had faith. Finally, as far as the molecular properties of material bodies are concerned, the inclusion seemed easier, but the details were very hazy. In a word, hopes were vast and strong, but vague.

To-day, what do we see? In the first place, a step in advance — immense progress. The relations between light and electricity are now known; the three domains of light, electricity, and magnetism, formerly separated, are now one; and this annexation seems definitive.

Nevertheless the conquest has caused us some sacrifices. Optical phenomena become particular cases in electric phenomena; as long as the former remained isolated, it was easy to explain them by movements which were thought to be known in all their details. That was easy enough; but any explanation to be accepted must now cover the whole domain of electricity. This cannot be done without difficulty.

The most satisfactory theory is that of Lorentz; it is unquestionably the theory that best explains the known facts, the one that throws into relief the greatest number of known relations, the one in which we find most traces of definitive construction. That it still possesses a serious fault I have shown above. It is in contradiction with Newton's law that action and re-action are equal and opposite — or rather, this principle according to Lorentz cannot be applicable to matter alone; if it be true, it must take into account the action of the ether on matter, and the re-action of the matter on the ether. Now, in the new order, it is very likely that things do not happen in this way.

However this may be, it is due to Lorentz that the results of Fizeau on the optics of moving bodies, the laws of normal and abnormal dispersion and of absorption are connected with each other and with the other properties of the ether, by bonds which no doubt will not be readily severed. Look at the ease with which the new Zeeman phenomenon found its place, and even aided the classification of Faraday's magnetic rotation, which had defied all Maxwell's efforts. This facility proves that Lorentz's theory is not a mere artificial combination which must eventually find its solvent. It will probably have to be modified, but not destroyed.

The only object of Lorentz was to include in a single whole all the optics and electro-dynamics of moving bodies; he did not claim to give a mechanical explanation. Larmor goes further; keeping the essential part of Lorentz's theory, he grafts upon it, so to speak, MacCullagh's ideas on the direction of the movement of the ether. MacCullagh held that the velocity of the ether is the same in magnitude and direction as the magnetic force. Ingenious as is this attempt, the fault in Lorentz's theory remains, and is even aggravated. According to Lorentz, we do not know what the movements of the ether are; and because we do not know this, we may suppose them to be movements compensating those of matter, and re-affirming that action and re-action are equal and opposite. According to Larmor we know the movements of the ether, and we can prove that the compensation does not take place.

If Larmor has failed, as in my opinion he has, does it necessarily follow that a mechanical explanation is impossible? Far from it. I said above that as long as a phenomenon obeys the two principles of energy and least action, so long it allows of an unlimited number of mechanical explanations. And so with the phenomena of optics and electricity.

But this is not enough. For a mechanical explanation to be good it must be simple; to choose it from among all the explanations that are possible there must be other reasons than the necessity of making a choice. Well, we have no theory as yet which will satisfy this condition and consequently be of any use. Are we then to complain? That would be to forget the end we seek, which is not the mechanism; the true and only aim is unity.

We ought therefore to set some limits to our ambition. Let us not seek to formulate a mechanical explanation; let us be content to show that we can always find one if we wish. In this we have succeeded. The principle of the conservation of energy has always been confirmed, and now it has a fellow in the principle of least action, stated in the form appropriate to physics. This has also been verified, at least as far as concerns the reversible phenomena which obey Lagrange's equations — in other words, which obey the most general laws of physics. The irreversible phenomena are much more difficult to bring into line; but they, too, are being co-ordinated and tend to come into the unity. The light which illuminates them comes from Carnot's principle. For a long time thermo-dynamics was confined to the study of the dilatations of bodies and of their change of state. For some time past it has been growing bolder, and has considerably extended its domain. We owe to it the theories of the voltaic cell and of their thermo-electric phenomena; there is not a corner in physics which it has not explored, and it has even attacked chemistry itself. The same laws hold good; everywhere, disguised in some form or other, we find Carnot's principle; everywhere also appears that eminently abstract concept of entropy which is as universal as the concept of energy, and like it, seems to conceal a reality. It seemed that radiant heat must escape, but recently that, too, has been brought under the same laws.

In this way fresh analogies are revealed which may be often pursued in detail; electric resistance resembles the viscosity of fluids; hysteresis would rather be like the friction of solids. In all cases friction appears to be the type most imitated by the most diverse irreversible phenomena, and this relationship is real and profound.

A strictly mechanical explanation of these phenomena has also been sought, but, owing to their nature, it is hardly likely that it will be found. To find it, it has been necessary to suppose that the irreversibility is but apparent, that the elementary phenomena are reversible and obey the known laws of dynamics. But the elements are extremely numerous, and become blended more and more, so that to our crude sight all appears to tend towards uniformity — *i.e.*, all seems to progress in the same direction, and that without hope of return. The apparent irreversibility is therefore but an effect of the law of great numbers. Only a being of infinitely subtle senses, such as Maxwell's demon, could unravel this tangled skein and turn back the course of the universe.

This conception, which is connected with the kinetic theory of gases, has cost great effort and has not, on the whole, been fruitful; it may become so. This is not the place to examine if it leads to contradictions, and if it is in conformity with the true nature of things.

Let us notice, however, the original ideas of M. Gouy on the Brownian movement. According to this scientist, this singular movement does not obey Carnot's principle. The particles which it sets moving would be smaller than the meshes of that tightly drawn net; they would thus be ready to separate them, and thereby to set back the course of the universe. One can almost see Maxwell's demon at work.[1]

To resume, phenomena long known are gradually being better classified, but new phenomena come to claim their place, and most of them, like the Zeeman effect, find it at once. Then we have the cathode rays, the X-rays, uranium and radium rays; in fact, a whole world of which none had suspected the existence. How many unexpected guests to find a place for! No one can yet predict the place they will occupy, but I do not believe they will destroy the general unity; I think that they will rather complete it. On the one hand, indeed, the new radiations seem to be connected with the phenomena of luminosity; not only do they excite fluorescence, but they sometimes come into existence under the same conditions as that property; neither are they unrelated to the cause which produces the electric spark under the action of ultra-violet light. Finally, and most important of all, it is believed that in all these phenomena there exist ions, animated, it is true, with velocities far greater than those of electrolytes. All this is very vague, but it will all become clearer.

Phosphorescence and the action of light on the spark were regions rather isolated, and consequently somewhat neglected by investigators. It is to be hoped that a new path will now be made which will facilitate their communications with the rest of science. Not only do we discover new phenomena, but those we think we know are revealed in unlooked-for aspects. In the free ether the laws preserve their majestic simplicity, but matter properly so called seems more and more complex; all we can say of it is but approximate, and our formulae are constantly requiring new terms.

But the ranks are unbroken, the relations that we have discovered between objects we thought simple still hold good between the same objects when their complexity is recognised, and that alone is the important thing. Our equations become, it is true, more and more complicated, so as to embrace more closely the complexity of nature; but nothing is changed in the relations which enable these equations to be derived from each other. In a word, the form of these equations persists. Take for instance the laws of reflection. Fresnel established them by a simple and attractive theory which experiment seemed to confirm. Subsequently, more accurate researches have shown that this verification was but approximate; traces of elliptic polarisation were detected everywhere. But it is owing to the first approximation that the cause of these anomalies was found in the existence of a transition layer, and all the essentials of Fresnel's theory have remained. We cannot help reflecting that all these relations would never have been noted if there had been doubt in the first place as to the complexity of the objects they connect. Long ago it was said: If Tycho had had instruments ten times as precise, we would never have had a Kepler, or a Newton, or Astronomy. It is a misfortune for a science to be born too late, when the means of observation have become too perfect. That is what is happening at this moment with respect to physical chemistry; the founders are hampered in their general grasp by third and fourth decimal places; happily they are men of robust faith. As we get to know the properties of matter better we see that continuity reigns. From the work of Andrews and Van der Waals, we see how the transition from the liquid to the gaseous state is made, and that it is not abrupt. Similarly, there is no gap between the liquid and solid states, and in the proceedings of a recent Congress we see memoirs on the rigidity of liquids side by side with papers on the flow of solids.

With this tendency there is no doubt a loss of simplicity. Such and such an effect was represented by straight lines; it is now necessary to connect these lines by more or less complicated curves. On the other hand, unity is gained. Separate categories quieted but did not satisfy the mind.

Finally, a new domain, that of chemistry, has been invaded by the method of physics, and we see the birth of physical chemistry. It is still quite young, but already it has enabled us to connect such phenomena as electrolysis, osmosis, and the movements of ions.

From this cursory exposition what can we conclude? Taking all things into account, we have approached the realisation of unity. This has not been done as quickly as was hoped fifty years ago, and the path predicted has not always been followed; but, on the whole, much ground has been gained.

Footnotes

1. Clerk Maxwell imagined some supernatural agency at work, sorting molecules in a gas of uniform temperature into (a) those possessing kinetic energy above the average, (b) those possessing kinetic energy below the average. [Tr.]

CHAPTER XI.
THE CALCULUS OF PROBABILITIES.

No doubt the reader will be astonished to find reflections on the calculus of probabilities in such a volume as this. What has that calculus to do with physical science? The questions I shall raise — without, however, giving them a solution — are naturally raised by the philosopher who is examining the problems of physics. So far is this the case, that in the two preceding chapters I have several times used the words "probability" and "chance." " Predicted facts," as I said above, "can only be probable." However solidly founded a prediction may appear to be, we are never absolutely certain that experiment will not prove it false; but the probability is often so great that practically it may be accepted. And a little farther on I added: — "See what a part the belief in simplicity plays in our generalisations. We have verified a simple law in a large number of particular cases, and we refuse to admit that this so-often-repeated coincidence is a mere effect of chance." Thus, in a multitude of circumstances the physicist is often in the same position as the gambler who reckons up his chances. Every time that he reasons by induction, he more or less consciously requires the calculus of probabilities, and that is why I am obliged to open this chapter parenthetically, and to interrupt our discussion of method in the physical sciences in order to examine a little closer what this calculus is worth, and what dependence we may place upon it. The very name of the calculus of probabilities is a paradox. Probability as opposed to certainty is what one does not know, and how can we calculate the unknown? Yet many eminent scientists have devoted themselves to this calculus, and it cannot be denied that science has drawn there from no small advantage. How can we explain this apparent contradiction? Has probability been defined? Can it even be defined? And if it can not, how can we venture to reason upon it? The definition, it will be said, is very simple. The probability of an event is the ratio of the number of cases favourable to the event to the total number of possible cases. A simple example will show how incomplete this definition is:—I throw two dice. What is the probability that one of the two at least turns up a 6? Each can turn up in six different ways; the number of possible cases is 6 x 6 = 36. The number of favourable cases is 11; the probability is $\frac{11}{36}$. That is the correct solution. But why cannot we just as well proceed as follows?—The points which turn up on the two dice form $^{6 \times 7}/_2$ = 21 different combinations. Among these combinations, six are favourable; the probability is $\frac{6}{21}$. Now why is the first method of calculating the number of possible cases more legitimate than the second? In any case it is not the definition that tells us. We are therefore bound to complete the definition by saying, "... to the total number of possible cases, provided the cases are equally probable." So we are compelled to define the probable by the probable. How can we know that two possible cases are equally probable? Will it be by a convention? If we insert at the beginning of every problem an explicit convention, well and good! We then have nothing to do but to apply the rules of arithmetic and algebra, and we complete our calculation, when our result cannot be called in question. But if we wish to make the slightest application of this result, we must prove that our convention is legitimate, and we shall find ourselves in the presence of the very difficulty we thought we had avoided. It may be said that common-sense is enough to show us the convention that should be adopted. Alas! M. Bertrand has amused himself by discussing the following simple problem:— "What is the probability that a chord of a circle may be greater than the side of the inscribed equilateral triangle?" The illustrious geometer successively adopted two conventions which seemed to be equally imperative in the eyes of common-sense, and with one convention he finds ½, and with the other ⅓. The conclusion which seems to follow from this is that the calculus of probabilities is a useless science, that the obscure instinct which we call common-sense, and to which we appeal for the legitimisation of our conventions, must be distrusted. But to this conclusion we can no longer subscribe. We cannot do without that obscure instinct. Without it, science would be impossible, and without it we could neither discover nor apply a law. Have we any right, for instance, to enunciate Newton's law? No doubt numerous observations are in agreement with it, but is not that a simple fact of chance? and how do we know, besides, that this law which has been true for so many generations will not be untrue in the next? To this objection the only answer you can give is: It is very improbable. But grant the law. By means of it I can calculate the position of Jupiter in a year from now. Yet have I any right to say this? Who can tell if a gigantic mass of enormous velocity is not going to pass near the solar system and produce unforeseen perturbations? Here again the only answer is: It is very improbable. From this point of view all the sciences would only be unconscious applications of the calculus of probabilities. And if this calculus be condemned, then the whole of the sciences must also be condemned. I shall not dwell at length on scientific problems in which the intervention of the calculus of probabilities is more evident. In the forefront of these is the problem of interpolation, in which, knowing a certain number of values of a function, we try to discover the intermediary values. I may also mention the celebrated theory of errors of observation, to which I shall return later; the kinetic theory of gases, a well-known hypothesis wherein each gaseous molecule is supposed to describe an extremely complicated path, but in which, through the effect of great numbers, the mean phenomena which are all we observe obey the simple laws of Mariotte and Gay-Lussac. All these theories are based upon the laws of great numbers, and the calculus of probabilities would evidently involve them in its ruin. It is true that they have only a particular interest, and that, save as far as interpolation is concerned, they are sacrifices to which we might readily be resigned. But I have said above, it would not be these partial sacrifices that would be in question; it would be the legitimacy of the whole of science that would be challenged. I quite see that it might be said: We do not know, and yet we must act. As for action, we have not time to devote ourselves to an inquiry that will suffice to dispel our ignorance. Besides, such an inquiry would demand unlimited time. We must therefore make up our minds without knowing. This must be often done whatever may happen, and we must follow the rules although we may have but little confidence in them. What I know is, not that such a thing is true, but that the best course for me is to act as if it were true. The calculus of probabilities, and therefore science itself, would be no longer of any practical value.

Unfortunately the difficulty does not thus dis appear. A gambler wants to try a *coup*, and he asks my advice. If I give it him, I use the calculus of probabilities; but I shall not guarantee success. That is what I shall call *subjective probability*. In this case we might be content with the explanation of which I have just given a sketch. But assume that an observer is present at the play, that he knows of the *coup*, and that play goes on for a long time, and that he makes a summary of his notes. He will find that events have taken place in conformity with the laws of the calculus of probabilities. That is what I shall call *objective probability*, and it is this phenomenon which has to be explained. There are numerous Insurance Societies which apply the rules of the calculus of probabilities, and they distribute to their shareholders dividends, the objective reality of which cannot be contested. In order to explain them, we must do more than invoke our ignorance and the necessity of action. Thus, absolute scepticism is not admissible. We may distrust, but we cannot condemn *en bloc*. Discussion is necessary.

I. *Classification of the Problems of Probability*. — In order to classify the problems which are presented to us with reference to probabilities, we must look at them from different points of view, and first of all, from that of *generality*. I said above that probability is the ratio of the number of favourable to the number of possible cases. What for want of a better term I call generality will increase with the number of possible cases. This number may be finite, as, for instance, if we take a throw of the dice in which the number of possible cases is 36. That is the first degree of generality. But if we ask, for instance, what is the probability that a point within a circle is within the inscribed square, there are as many possible cases as there are points in the circle — that is to say, an infinite number. This is the second degree of generality. Generality can be pushed further still. We may ask the probability that a function will satisfy a given condition. There are then as many possible cases as one can imagine different functions. This is the third degree of generality, which we reach, for instance, when we try to find the most probable law after a finite number of observations. Yet we may place ourselves at a quite different point of view. If we were not ignorant there would be no probability, there could only be certainty. But our ignorance cannot be absolute, for then there would be no longer any probability at all. Thus the problems of probability may be classed according to the greater or less depth of this ignorance. In mathematics we may set ourselves problems in probability. What is the probability that the fifth decimal of a logarithm taken at random from a table is a 9. There is no hesitation in answering that this probability is 1-10th. Here we possess all the data of the problem. We can calculate our logarithm without having recourse to the table, but we need not give ourselves the trouble. This is the first degree of ignorance. In the physical sciences our ignorance is already greater. The state of a system at a given moment depends on two things — its initial state, and the law according to which that state varies. If we know both this law and this initial state, we have a simple mathematical problem to solve, and we fall back upon our first degree of ignorance. Then it often happens that we know the law and do not know the initial state. It may be asked, for instance, what is the present distribution of the minor planets? We know that from all time they have obeyed the laws of Kepler, but we do not know what was their initial distribution. In the kinetic theory of gases we assume that the gaseous molecules follow rectilinear paths and obey the laws of impact and elastic bodies; yet as we know nothing of their initial velocities, we know nothing of their present velocities. The calculus of probabilities alone enables us to predict the mean phenomena which will result from a combination of these velocities. This is the second degree of ignorance. Finally it is possible, that not only the initial conditions but the laws themselves are unknown. We then reach the third degree of ignorance, and in general we can no longer affirm anything at all as to the probability of a phenomenon. It often happens that instead of trying to discover an event by means of a more or less imperfect knowledge of the law, the events may be known, and we want to find the law; or that, instead of deducing effects from causes, we wish to deduce the causes from the effects. Now, these problems are classified as *probability of causes*, and are the most interesting of all from their scientific applications. I play at *écarté* with a gentleman whom I know to be perfectly honest. What is the chance that he turns up the king? It is 1/8. This is a problem of the probability of effects. I play with a gentleman whom I do not know. He has dealt ten times, and he has turned the king up six times. What is the chance that he is a sharper? This is a problem in the probability of causes. It may be said that it is the essential problem of the experimental method. I have observed *n* values of *x* and the corresponding values of *y*. I have found that the ratio of the latter to the former is practically constant. There is the event; what is the cause? Is it probable that there is a general law according to which *y* would be proportional to *x*, and that small divergencies are due to errors of observation? This is the type of question that we are ever asking, and which we unconsciously solve whenever we are engaged in scientific work. I am now going to pass in review these different categories of problems by discussing in succession what I have called subjective and objective probability.

II. *Probability in Mathematics.* — The impossibility of squaring the circle was shown in 1885, but before that date all geometers considered this impossibility as so "probable" that the Académie des Sciences rejected without examination the, alas! too numerous memoirs on this subject that a few unhappy madmen sent in every year. Was the Académie wrong? Evidently not, and it knew perfectly well that by acting in this manner it did not run the least risk of stifling a discovery of moment. The Académie could not have proved that it was right, but it knew quite well that its instinct did not deceive it. If you had asked the Academicians, they would have answered: "We have compared the probability that an unknown scientist should have found out what has been vainly sought for so long, with the probability that there is one madman the more on the earth, and the latter has appeared to us the greater." These are very good reasons, but there is nothing mathematical about them; they are purely psychological. If you had pressed them further, they would have added: "Why do you expect a particular value of a transcendental function to be an algebraical number; if π be the root of an algebraical equation, why do you expect this root to be a period of the function *sin 2x*, and why is it not the same with the other roots of the same equation?" To sum up, they would have invoked the principle of sufficient reason in its vaguest form. Yet what information could they draw from it? At most a rule of conduct for the employment of their time, which would be more usefully spent at their ordinary work than in reading a lucubration that inspired in them a legitimate distrust. But what I called above objective probability has nothing in common with this first problem. It is otherwise with the second. Let us consider the first 10,000 logarithms that we find in a table. Among these 10,000 logarithms I take one at random. What is the probability that its third decimal is an even number? You will say with out any hesitation that the probability is ½, and in fact if you pick out in a table the third decimals in these 10,000 numbers you will find nearly as many even digits as odd. Or, if you prefer it, let us write 10,000 numbers corresponding to our 10,000 logarithms, writing down for each of these numbers +1 if the third decimal of the corresponding logarithm is even, and -1 if odd; and then let us take the mean of these 10,000 numbers. I do not hesitate to say that the mean of these 10,000 units is probably zero, and if I were to calculate it practically, I would verify that it is extremely small. But this verification is needless. I might have rigorously proved that this mean is smaller than 0.003. To prove this result I should have had to make a rather long calculation for which there is no room here, and for which I may refer the reader to an article that I published in the *Revue générale des Sciences*, April 15th, 1899. The only point to which I wish to draw attention is the following. In this calculation I had occasion to rest my case on only two facts namely, that the first and second derivatives of the logarithm remain, in the interval considered, between certain limits. Hence our first conclusion is that the property is not only true of the logarithm but of any continuous function whatever, since the derivatives of every continuous function are limited. If I was certain beforehand of the result, it is because I have often observed analogous facts for other continuous functions; and next, it is because I went through in my mind in a more or less unconscious and imperfect manner the reasoning which led me to the preceding in equalities, just as a skilled calculator before finishing his multiplication takes into account what it ought to come to approximately. And besides, since what I call my intuition was only an incomplete summary of a piece of true reasoning, it is clear that observation has confirmed my predictions, and that the objective and subjective probabilities are in agreement. As a third example I shall choose the following: T— he number *u* is taken at random and *n* is a given very large integer. What is the mean value of sin *nu*? This problem has no meaning by itself. To give it one, a convention is required — namely, we agree that the probability for the number *u* to lie between *a* and *a + da* is *Φ(a)da;* that it is therefore proportional to the infinitely small interval *da*, and is equal to this multiplied by a function *Φ(a)*, only depending on *a*. As for this function I choose it arbitrarily, but I must assume it to be continuous. The value of sin *nu* remaining the same when *u* increases by 2π, I may without loss of generality assume that *u* lies between 0 and 2π, and I shall thus be led to suppose that Φ(a) is a periodic function whose period is 2π. The mean value that we seek is readily expressed by a simple integral, and it is easy to show that this integral is smaller than $\dfrac{2\pi M_K}{n^K}$, M_K being the maximum value of the Kth derivative of *Φ(u)*. We see then that if the

Kth derivative is finite, our mean value will tend towards zero when n increases indefinitely, and that more rapidly than $\dfrac{1}{n^{K-1}}$. The mean value of sin nu when n is very large is therefore zero. To define this value I required a convention, but the result remains the same *whatever that convention may be*. I have imposed upon myself but slight restrictions when I assumed that the function $\Phi(a)$ is continuous and periodic, and these hypotheses are so natural that we may ask ourselves how they can be escaped. Examination of the three preceding examples, so different in all respects, has already given us a glimpse on the one hand of the rôle of what philosophers call the principle of sufficient reason, and on the other hand of the importance of the fact that certain properties are common to all continuous functions. The study of probability in the physical sciences will lead us to the same result.

III. *Probability in the Physical Sciences.* — We now come to the problems which are connected with what I have called the second degree of ignorance — namely, those in which we know the law but do not know the initial state of the system. I could multiply examples, but I shall take only one. What is the probable present distribution of the minor planets on the zodiac? We know they obey the laws of Kepler. We may even, without changing the nature of the problem, suppose that their orbits are circular and situated in the same plane, a plane which we are given. On the other hand, we know absolutely nothing about their initial distribution. However, we do not hesitate to affirm that this distribution is now nearly uniform. Why? Let b be the longitude of a minor planet in the initial epoch that is to say, the epoch zero. Let a be its mean motion. Its longitude at the present time — *i.e.*, at the time t will be $at + b$. To say that the present distribution is uniform is to say that the mean value of the sines and cosines of multiples of $at + b$ is zero. Why do we assert this? Let us represent our minor planet by a point in a plane — namely, the point whose co-ordinates are a and b. All these representative points will be contained in a certain region of the plane, but as they are very numerous this region will appear dotted with points. We know nothing else about the distribution of the points. Now what do we do when we apply the calculus of probabilities to such a question as this? What is the probability that one or more representative points may be found in a certain portion of the plane? In our ignorance we are compelled to make an arbitrary hypothesis. To explain the nature of this hypothesis I may be allowed to use, instead of a mathematical formula, a crude but concrete image. Let us suppose that over the surface of our plane has been spread imaginary matter, the density of which is variable, but varies continuously. We shall then agree to say that the probable number of representative points to be found on a certain portion of the plane is proportional to the quantity of this imaginary matter which is found there. If there are, then, two regions of the plane of the same extent, the probabilities that a representative point of one of our minor planets is in one or other of these regions will be as the mean densities of the imaginary matter in one or other of the regions. Here then are two distributions, one real, in which the representative points are very numerous, very close together, but discrete like the molecules of matter in the atomic hypothesis; the other remote from reality, in which our representative points are replaced by imaginary continuous matter. We know that the latter cannot be real, but we are forced to adopt it through our ignorance. If, again, we had some idea of the real distribution of the representative points, we could arrange it so that in a region of some extent the density of this imaginary continuous matter may be nearly proportional to the number of representative points, or, if it is preferred, to the number of atoms which are contained in that region. Even that is impossible, and our ignorance is so great that we are forced to choose arbitrarily the function which defines the density of our imaginary matter. We shall be compelled to adopt a hypothesis from which we can hardly get away; we shall suppose that this function is continuous. That is sufficient, as we shall see, to enable us to reach our conclusion.

What is at the instant *t* the probable distribution of the minor planets — or rather, what is the mean value of the sine of the longitude at the moment *t* — *i.e.*, of sin (*at* + *b*)? We made at the outset an arbitrary convention, but if we adopt it, this probable value is entirely defined. Let us decompose the plane into elements of surface. Consider the value of sin (*at* + *b*) at the centre of each of these elements. Multiply this value by the surface of the element and by the corresponding density of the imaginary matter. Let us then take the sum for all the elements of the plane. This sum, by definition, will be the probable mean value we seek, which will thus be expressed by a double integral. It may be thought at first that this mean value depends on the choice of the function Φ which defines the density of the imaginary matter, and as this function Φ is arbitrary, we can, according to the arbitrary choice which we make, obtain a certain mean value. But this is not the case. A simple calculation shows us that our double integral decreases very rapidly as *t* increases. Thus, I cannot tell what hypothesis to make as to the probability of this or that initial distribution, but when once the hypothesis is made the result will be the same, and this gets me out of my difficulty. Whatever the function Φ may be, the mean value tends towards zero as *t* increases, and as the minor planets have certainly accomplished a very large number of revolutions, I may assert that this mean value is very small. I may give to Φ any value I choose, with one restriction: this function must be continuous; and, in fact, from the point of view of subjective probability, the choice of a discontinuous function would have been unreasonable. What reason could I have, for instance, for supposing that the initial longitude might be exactly o°, but that it could not lie between o° and 1°?

The difficulty reappears if we look at it from the point of view of objective probability; if we pass from our imaginary distribution in which the supposititious matter was assumed to be continuous, to the real distribution in which our representative points are formed as discrete atoms. The mean value of sin (*at* + *b*) will be represented quite simply by

$$\frac{1}{n} \sum \sin\ (at + b),$$

n being the number of minor planets. Instead of a double integral referring to a continuous function, we shall have a sum of discrete terms. However, no one will seriously doubt that this mean value is practically very small. Our representative points being very close together, our discrete sum will in general differ very little from an integral. An integral is the limit towards which a sum of terms tends when the number of these terms is indefinitely increased. If the terms are very numerous, the sum will differ very little from its limit—that is to say, from the integral, and what I said of the latter will still be true of the sum itself. But there are exceptions. If, for instance, for all the minor planets *b* = π/2 -*at*, the longitude of all the planets at the time *t* would be π/2, and the mean value in question would be evidently unity. For this to be the case at the time o, the minor planets must have all been lying on a kind of spiral of peculiar form, with its spires very close together. All will admit that such an initial distribution is extremely improbable (and even if it were realised, the distribution would not be uniform at the present time—for example, on the 1st January 1900; but it would become so a few years later). Why, then, do we think this initial distribution improbable? This must be explained, for if we are wrong in rejecting as improbable this absurd hypothesis, our inquiry breaks down, and we can no longer affirm any thing on the subject of the probability of this or that present distribution. Once more we shall invoke the principle of sufficient reason, to which we must always recur. We might admit that at the beginning the planets were distributed almost in a straight line. We might admit that they were irregularly distributed. But it seems to us that there is no sufficient reason for the unknown cause that gave them birth to have acted along a curve so regular and yet so complicated, which would appear to have been expressly chosen so that the distribution at the present day would not be uniform.

IV. *Rouge et Noir.* — The questions raised by games of chance, such as roulette, are, fundamentally, quite analogous to those we have just treated. For example, a wheel is divided into thirty-seven equal compartments, alternately red and black. A ball is spun round the wheel, and after having moved round a number of times, it stops in front of one of these sub-divisions. The probability that the division is red is obviously ½. The needle describes an angle θ, including several complete revolutions. I do not know what is the probability that the ball is spun with such a force that this angle should lie between θ and θ+dθ, but I can make a convention. I can suppose that this probability is Φ(θ)dθ. As for the function Φ(θ), I can choose it in an entirely arbitrary manner. I have nothing to guide me in my choice, but I am naturally induced to suppose the function to be continuous. Let ε be a length (measured on the circumference of the circle of radius unity) of each red and black compartment. We have to calculate the integral of Φ(θ)dθ, extending it on the one hand to all the red, and on the other hand to all the black compartments, and to compare the results. Consider an interval 2ε comprising two consecutive red and black compartments. Let M and m be the maximum and minimum values of the function Φ(θ) in this interval. The integral extended to the red compartments will be smaller than Σmε; extended to the black it will be greater than Σmε. The difference will therefore be smaller than Σ(M-m)ε. But if the function Φ is supposed continuous, and if on the other hand the interval ε is very small with respect to the total angle described by the needle, the difference M-m will be very small. The difference of the two integrals will be therefore very small, and the probability will be very nearly ½. We see that without knowing anything of the function Φ we must act as if the probability were ½. And on the other hand it explains why, from the objective point of view, if I watch a certain number of *coups*, observation will give me almost as many black *coups* as red. All the players know this objective law; but it leads them into a remarkable error, which has often been exposed, but into which they are always falling. When the red has won, for example, six times running, they bet on black, thinking that they are playing an absolutely safe game, because they say it is a very rare thing for the red to win seven times running. In reality their probability of winning is still ½. Observation shows, it is true, that the series of seven consecutive reds is very rare, but series of six reds followed by a black are also very rare. They have noticed the rarity of the series of seven reds; if they have not remarked the rarity of six reds and a black, it is only because such series strike the attention less.

V. *The Probability of Causes.* — We now come to the problems of the probability of causes, the most important from the point of view of scientific applications. Two stars, for instance, are very close together on the celestial sphere. Is this apparent contiguity a mere effect of chance? Are these stars, although almost on the same visual ray, situated at very different distances from the earth, and therefore very far indeed from one another? or does the apparent correspond to a real contiguity? This is a problem on the probability of causes.

First of all, I recall that at the outset of all problems of probability of effects that have occupied our attention up to now, we have had to use a convention which was more or less justified; and if in most cases the result was to a certain extent independent of this convention, it was only the condition of certain hypotheses which enabled us à *priori* to reject discontinuous functions, for example, or certain absurd conventions. We shall again find something analogous to this when we deal with the probability of causes. An effect may be produced by the cause *a* or by the cause *b*. The effect has just been observed. We ask the probability that it is due to the cause *a*. This is an à *posteriori* probability of cause. But I could not calculate it, if a convention more or less justified did not tell me in advance what is the à *priori* probability for the cause *a* to come into play — I mean the probability of this event to some one who had not observed the effect. To make my meaning clearer, I go back to the game of *écarté* mentioned before. My adversary deals for the first time and turns up a king. What is the probability that he is a sharper? The formulae ordinarily taught give 8/9, a result which is obviously rather surprising. If we look at it closer, we see that the conclusion is arrived at as if, before sitting down at the table, I had considered that there was one chance in two that my adversary was not honest. An absurd hypothesis, because in that case I should certainly not have played with him; and this explains the absurdity of the conclusion. The function on the à priori probability was unjustified, and that is why the conclusion of the à posteriori probability led me into an inadmissible result. The importance of this preliminary convention is obvious. I shall even add that if none were made, the problem of the à posteriori probability would have no meaning. It must be always made either explicitly or tacitly.

Let us pass on to an example of a more scientific character. I require to determine an experimental law; this law, when discovered, can be represented by a curve. I make a certain number of isolated observations, each of which may be represented by a point. When I have obtained these different points, I draw a curve between them as carefully as possible, giving my curve a regular form, avoiding sharp angles, accentuated inflexions, and any sudden variation of the radius of curvature. This curve will represent to me the probable law, and not only will it give me the values of the functions intermediary to those which have been observed, but it also gives me the observed values more accurately than direct observation does; that is why I make the curve pass near the points and not through the points themselves.

Here, then, is a problem in the probability of causes. The effects are the measurements I have recorded; they depend on the combination of two causes — the true law of the phenomenon and errors of observation. Knowing the effects, we have to find the probability that the phenomenon shall obey this law or that, and that the observations have been accompanied by this or that error. The most probable law, therefore, corresponds to the curve we have traced, and the most probable error is represented by the distance of the corresponding point from that curve. But the problem has no meaning if before the observations I had an à priori idea of the probability of this law or that, or of the chances of error to which I am exposed. If my instruments are good (and I knew whether this is so or not before beginning the observations), I shall not draw the curve far from the points which represent the rough measurements. If they are inferior, I may draw it a little farther from the points, so that I may get a less sinuous curve; much will be sacrificed to regularity.

Why, then, do I draw a curve without sinuosities? Because I consider à priori a law represented by a continuous function (or function the derivatives of which to a high order are small), as more probable than a law not satisfying those conditions. But for this conviction the problem would have no meaning; interpolation would be impossible; no law could be deduced from a finite number of observations; science would cease to exist.

Fifty years ago physicists considered, other things being equal, a simple law as more probable than a complicated law. This principle was even invoked in favour of Mariotte's law as against that of Regnault. But this belief is now repudiated; and yet, how many times are we compelled to act as though we still held it! However that may be, what remains of this tendency is the belief in continuity, and as we have just seen, if the belief in continuity were to disappear, experimental science would become impossible.

VI. *The Theory of Errors.* — We are thus brought to consider the theory of errors which is directly connected with the problem of the probability of causes. Here again we find *effects* — to wit, a certain number of irreconcilable observations, and we try to find the *causes* which are, on the one hand, the true value of the quantity to be measured, and, on the other, the error made in each isolated observation. We must calculate the probable a *posteriori* value of each error, and therefore the probable value of the quantity to be measured. But, as I have just explained, we cannot undertake this calculation unless we admit à *priori* — i.e., before any observations are made — that there is a law of the probability of errors. Is there a law of errors? The law to which all calculators assent is Gauss's law, that is represented by a certain transcendental curve known as the "bell."

But it is first of all necessary to recall the classic distinction between systematic and accidental errors. If the metre with which we measure a length is too long, the number we get will be too small, and it will be no use to measure several times — that is a systematic error. If we measure with an accurate metre, we may make a mistake, and find the length sometimes too large and sometimes too small, and when we take the mean of a large number of measurements, the error will tend to grow small. These are accidental errors.

It is clear that systematic errors do not satisfy Gauss's law, but do accidental errors satisfy it? Numerous proofs have been attempted, almost all of them crude paralogisms. But starting from the following hypotheses we may prove Gauss's law: the error is the result of a very large number of partial and independent errors; each partial error is very small and obeys any law of probability whatever, provided the probability of a positive error is the same as that of an equal negative error. It is clear that these conditions will be often, but not always, fulfilled, and we may reserve the name of accidental for errors which satisfy them.

We see that the method of least squares is not legitimate in every case; in general, physicists are more distrustful of it than astronomers. This is no doubt because the latter, apart from the systematic errors to which they and the physicists are subject alike, have to contend with an extremely important source of error which is entirely accidental — I mean atmospheric undulations. So it is very curious to hear a discussion between a physicist and an astronomer about a method of observation. The physicist, persuaded that one good measurement is worth more than many bad ones, is pre-eminently concerned with the elimination by means of every precaution of the final systematic errors; the astronomer retorts: "But you can only observe a small number of stars, and accidental errors will not disappear."

What conclusion must we draw? Must we continue to use the method of least squares? We must distinguish. We have eliminated all the systematic errors of which we have any suspicion; we are quite certain that there are others still, but we cannot detect them; and yet we must make up our minds and adopt a definitive value which will be regarded as the probable value; and for that purpose it is clear that the best thing we can do is to apply Gauss's law. We have only applied a practical rule referring to subjective probability. And there is no more to be said.

Yet we want to go farther and say that not only the probable value is so much, but that the probable error in the result is so much. *This is absolutely invalid*: it would be true only if we were sure that all the systematic errors were eliminated, and of that we know absolutely nothing. We have two series of observations; by applying the law of least squares we find that the probable error in the first series is twice as small as in the second. The second series may, how ever, be more accurate than the first, because the first is perhaps affected by a large systematic error. All that we can say is, that the first series is *probably* better than the second because its accidental error is smaller, and that we have no reason for affirming that the systematic error is greater for one of the series than for the other, our ignorance on this point being absolute.

VII. *Conclusions*. In the preceding lines I have set several problems, and have given no solution. I do not regret this, for perhaps they will invite the reader to reflect on these delicate questions.

However that may be, there are certain points which seem to be well established. To undertake the calculation of any probability, and even for that calculation to have any meaning at all, we must admit, as a point of departure, an hypothesis or convention which has always something arbitrary about it. In the choice of this convention we can be guided only by the principle of sufficient reason.

Unfortunately, this principle is very vague and very elastic, and in the cursory examination we have just made we have seen it assume different forms. The form under which we meet it most often is the belief in continuity, a belief which it would be difficult to justify by apodeictic reasoning, but without which all science would be impossible. Finally, the problems to which the calculus of probabilities may be applied with profit are those in which the result is independent of the hypothesis made at the outset, provided only that this hypothesis satisfies the condition of continuity.

Part IV: Nature

CHAPTER XII.[1]
OPTICS AND ELECTRICITY.

Fresnel's Theory. — The best example that can be chosen is the theory of light and its relations to the theory of electricity. It is owing to Fresnel that the science of optics is more advanced than any other branch of physics. The theory called the theory of undulations forms a complete whole, which is satisfying to the mind; but we must not ask from it what it cannot give us. The object of mathematical theories is not to reveal to us the real nature of things; that would be an unreasonable claim. Their only object is to co-ordinate the physical laws with which physical experiment makes us acquainted, the enunciation of which, without the aid of mathematics, we should be unable to effect. Whether the ether exists or not matters little — let us leave that to the metaphysicians; what is essential for us is, that everything happens as if it existed, and that this hypothesis is found to be suitable for the explanation of phenomena. After all, have we any other reason for believing in the existence of material objects? That, too, is only a convenient hypothesis; only, it will never cease to be so, while some day, no doubt, the ether will be thrown aside as useless.

But at the present moment the laws of optics, and the equations which translate them into the language of analysis, hold good — at least as a first approximation. It will therefore be always useful to study a theory which brings these equations into connection.

The undulatory theory is based on a molecular hypothesis; this is an advantage to those who think they can discover the cause under the law. But others find in it a reason for distrust; and this distrust seems to me as unfounded as the illusions of the former. These hypotheses play but a secondary role. They may be sacrificed, and the sole reason why this is not generally done is, that it would involve a certain loss of lucidity in the explanation. In fact, if we look at it a little closer we shall see that we borrow from molecular hypotheses but two things — the principle of the conservation of energy, and the linear form of the equations, which is the general law of small movements as of all small variations. This explains why most of the conclusions of Fresnel remain unchanged when we adopt the electro magnetic theory of light.

Maxwell's Theory. — We all know that it was Maxwell who connected by a slender tie two branches of physics — optics and electricity — until then unsuspected of having anything in common. Thus blended in a larger aggregate, in a higher harmony, Fresnel's theory of optics did not perish. Parts of it are yet alive, and their mutual relations are still the same. Only, the language which we use to express them has changed; and, on the other hand, Maxwell has revealed to us other relations, hitherto unsuspected, between the different branches of optics and the domain of electricity.

The first time a French reader opens Maxwell's book, his admiration is tempered with a feeling of uneasiness, and often of distrust.

It is only after prolonged study, and at the cost of much effort, that this feeling disappears. Some minds of high calibre never lose this feeling. Why is it so difficult for the ideas of this English scientist to become acclimatised among us? No doubt the education received by most enlightened Frenchmen predisposes them to appreciate precision and logic more than any other qualities. In this respect the old theories of mathematical physics gave us complete satisfaction. All our masters, from Laplace to Cauchy, proceeded along the same lines. Starting with clearly enunciated hypotheses, they deduced from them all their consequences with mathematical rigour, and then compared them with experiment. It seemed to be their aim to give to each of the branches of physics the same precision as to celestial mechanics. A mind accustomed to admire such models is not easily satisfied with a theory. Not only will it not tolerate the least appearance of contradiction, but it will expect the different parts to be logically connected with one another, and will require the number of hypotheses to be reduced to a minimum.

This is not all; there will be other demands which appear to me to be less reasonable. Behind the matter of which our senses are aware, and which is made known to us by experiment, such a thinker will expect to see another kind of matter — the only true matter in its opinion — which will no longer have anything but purely geometrical qualities, and the atoms of which will be mathematical points subject to the laws of dynamics alone. And yet he will try to represent to himself, by an unconscious contradiction, these invisible and colourless atoms, and therefore to bring them as close as possible to ordinary matter.

Then only will he be thoroughly satisfied, and he will then imagine that he has penetrated the secret of the universe. Even if the satisfaction is fallacious, it is none the less difficult to give it up. Thus, on opening the pages of Maxwell, a French man expects to find a theoretical whole, as logical and as precise as the physical optics that is founded on the hypothesis of the ether. He is thus preparing for himself a disappointment which I should like the reader to avoid; so I will warn him at once of what he will find and what he will not find in Maxwell.

Maxwell does not give a mechanical explanation of electricity and magnetism; he confines himself to showing that such an explanation is possible. He shows that the phenomena of optics are only a particular case of electro-magnetic phenomena. From the whole theory of electricity a theory of light can be immediately deduced. Unfortunately the converse is not true; it is not always easy to find a complete explanation of electrical phenomena. In particular it is not easy if we take as our starting-point Fresnel's theory; to do so, no doubt, would be impossible; but none the less we must ask ourselves if we are compelled to surrender admirable results which we thought we had definitively acquired. That seems a step backwards, and many sound intellects will not willingly allow of this.

Should the reader consent to set some bounds to his hopes, he will still come across other difficulties. The English scientist does not try to erect a unique, definitive, and well-arranged building; he seems to raise rather a large number of provisional and independent constructions, between which communication is difficult and sometimes impossible. Take, for instance, the chapter in which electrostatic attractions are explained by the pressures and tensions of the dielectric medium. This chapter might be sup pressed without the rest of the book being thereby less clear or less complete, and yet it contains a theory which is self-sufficient, and which can be understood without reading a word of what precedes or follows. But it is not only independent of the rest of the book; it is difficult to reconcile it with the fundamental ideas of the volume. Maxwell does not even attempt to reconcile it; he merely says: "I have not been able to make the next step — namely, to account by mechanical considerations for these stresses in the dielectric."

This example will be sufficient to show what I mean; I could quote many others. Thus, who would suspect on reading the pages devoted to magnetic rotatory polarisation that there is an identity between optical and magnetic phenomena?

We must not flatter ourselves that we have avoided every contradiction, but we ought to make up our minds. Two contradictory theories, provided that they are kept from overlapping, and that we do not look to find in them the explanation of things, may, in fact, be very useful instruments of research; and perhaps the reading of Maxwell would be less suggestive if he had not opened up to us so many new and divergent ways. But the fundamental idea is masked, as it were. So far is this the case, that in most works that are popularised, this idea is the only point which is left completely untouched. To show the importance of this, I think I ought to explain in what this fundamental idea consists; but for that purpose a short digression is necessary.

The Mechanical Explanation of Physical Phenomena. — In every physical phenomenon there is a certain number of parameters which are reached directly by experiment, and which can be measured. I shall call them the parameters q. Observation next teaches us the laws of the variations of these parameters, and these laws can be generally stated in the form of differential equations which connect together the parameters q and time. What can be done to give a mechanical interpretation to such a phenomenon? We may endeavour to explain it, either by the movements of ordinary matter, or by those of one or more hypothetical fluids. These fluids will be considered as formed of a very large number of isolated molecules m. When may we say that we have a complete mechanical explanation of the phenomenon? It will be, on the one hand, when we know the differential equations which are satisfied by the co-ordinates of these hypothetical molecules m, equations which must, in addition, conform to the laws of dynamics; and, on the other hand, when we know the relations which define the co-ordinates of the molecules m as functions of the parameters q, attainable by experiment. These equations, as I have said, should conform to the principles of dynamics, and, in particular, to the principle of the conservation of energy, and to that of least action. The first of these two principles teaches us that the total energy is constant, and may be divided into two parts:

(1) Kinetic energy, or *vis viva*, which depends on the masses of the hypothetical molecules m, and on their velocities. This I shall call T. (2) The potential energy which depends only on the co-ordinates of these molecules, and this I shall call U. It is the sum of the energies T and U that is constant.

Now what are we taught by the principle of least action? It teaches us that to pass from the initial position occupied at the instant t_0 to the final position occupied at the instant t_1, the system must describe such a path that in the interval of time between the instant t_0 and t_1, the mean value of the action — *i.e.*, the difference between the two energies T and U, must be as small as possible. The first of these two principles is, moreover, a consequence of the second. If we know the functions T and U, this second principle is sufficient to determine the equations of motion.

Among the paths which enable us to pass from one position to another, there is clearly one for which the mean value of the action is smaller than for all the others. In addition, there is only such path; and it follows from this, that the principle of least action is sufficient to determine the path followed, and therefore the equations of motion. We thus obtain what are called the equations of Lagrange. In these equations the independent variables are the co-ordinates of the hypothetical molecules m; but I now assume that we take for the variables the parameters q, which are directly accessible to experiment.

The two parts of the energy should then be expressed as a function of the parameters q and their derivatives; it is clear that it is under this form that they will appear to the experimenter. The latter will naturally endeavour to define kinetic and potential energy by the aid of quantities he can directly observe.[2] If this be granted, the system will always proceed from one position to another by such a path that the mean value of the action is a minimum. It matters little that T and U are now expressed by the aid of the parameters q and their derivatives; it matters little that it is also by the aid of these parameters that we define the initial and final positions; the principle of least action will always remain true.

Now here again, of the whole of the paths which lead from one position to another, there is one and only one for which the mean action is a minimum. The principle of least action is therefore sufficient for the determination of the differential equations which define the variations of the parameters q. The equations thus obtained are another form of Lagrange's equations. To form these equations we need not know the relations which connect the parameters q with the co-ordinates of the hypothetical molecules, nor the masses of the molecules, nor the expression of U as a function of the co-ordinates of these molecules. All we need know is the expression of U as a function of the parameters q, and that of T as a function of the parameters q and their derivatives — *i.e.*, the expressions of the kinetic and potential energy in terms of experimental data.

One of two things must now happen. Either for a convenient choice of T and U the Lagrangian equations, constructed as we have indicated, will be identical with the differential equations deduced from experiment, or there will be no functions T and U for which this identity takes place. In the latter case it is clear that no mechanical explanation is possible. The *necessary* condition for a mechanical explanation to be possible is therefore this: that we may choose the functions T and U so as to satisfy the principle of least action, and of the conservation of energy. Besides, this condition is *sufficient*. Suppose, in fact, that we have found a function U of the parameters q, which represents one of the parts of energy, and that the part of the energy which we represent by T is a function of the parameters q and their derivatives; that it is a polynomial of the second degree with respect to its derivatives, and finally that the Lagrangian equations formed by the aid of these two functions T and U are in conformity with the data of the experiment. How can we deduce from this a mechanical explanation? U must be regarded as the potential energy of a system of which T is the kinetic energy. There is no difficulty as far as U is concerned, but can T be regarded as the *vis viva* of a material system?

It is easily shown that this is always possible, and in an unlimited number of ways. I will be content with referring the reader to the pages of the preface of my *Électricité et Optique* for further details. Thus, if the principle of least action cannot be satisfied, no mechanical explanation is possible; if it can be satisfied, there is not only one explanation, but an unlimited number, whence it follows that since there is one there must be an unlimited number.

One more remark. Among the quantities that may be reached by experiment directly we shall consider some as the co-ordinates of our hypothetical molecules, some will be our parameters q, and the rest will be regarded as dependent not only on the co-ordinates but on the velocities — or what comes to the same thing, we look on them as derivatives of the parameters q, or as combinations of these parameters and their derivatives.

Here then a question occurs: among all these quantities measured experimentally which shall we choose to represent the parameters q? and which shall we prefer to regard as the derivatives of these parameters? This choice remains arbitrary to a large extent, but a mechanical explanation will be possible if it is done so as to satisfy the principle of least action.

Next, Maxwell asks: Can this choice and that of the two energies T and U be made so that electric phenomena will satisfy this principle? Experiment shows us that the energy of an electro-magnetic field decomposes into electro-static and electro-dynamic energy. Maxwell recognised that if we regard the former as the potential energy U, and the latter as the kinetic energy T, and that if on the other hand we take the electro-static charges of the conductors as the parameters q, and the intensity of the currents as derivatives of other parameters q — under these conditions, Maxwell has recognised that electric phenomena satisfies the principle of least action. He was then certain of a mechanical explanation. If he had expounded this theory at the beginning of his first volume, instead of relegating it to a corner of the second, it would not have escaped the attention of most readers. If therefore a phenomenon allows of a complete mechanical explanation, it allows of an unlimited number of others, which will equally take into account all the particulars revealed by experiment. And this is confirmed by the history of every branch of physics. In Optics, for instance, Fresnel believed vibration to be perpendicular to the plane of polarisation; Neumann holds that it is parallel to that plane. For a long time an *experimentum crucis* was sought for, which would enable us to decide between these two theories, but in vain. In the same way, without going out of the domain of electricity, we find that the theory of two fluids and the single fluid theory equally account in a satisfactory manner for all the laws of electro-statics. All these facts are easily explained, thanks to the properties of the Lagrange equations.

It is easy now to understand Maxwell's fundamental idea. To demonstrate the possibility of a mechanical explanation of electricity we need not trouble to find the explanation itself; we need only know the expression of the two functions T and U, which are the two parts of energy, and to form with these two functions Lagrange's equations, and then to compare these equations with the experimental laws.

How shall we choose from all the possible explanations one in which the help of experiment will be wanting? The day will perhaps come when physicists will no longer concern themselves with questions which are inaccessible to positive methods, and will leave them to the metaphysicians. That day has not yet come; man does not so easily resign himself to remaining for ever ignorant of the causes of things. Our choice cannot be therefore any longer guided by considerations in which personal appreciation plays too large a part. There are, however, solutions which all will reject because of their fantastic nature, and others which all will prefer because of their simplicity. As far as magnetism and electricity are concerned, Maxwell abstained from making any choice. It is not that he has a systematic contempt for all that positive methods cannot reach, as may be seen from the time he has devoted to the kinetic theory of gases. I may add that if in his *magnum opus* he develops no complete explanation, he has attempted one in an article in the *Philosophical Magazine*. The strangeness and the complexity of the hypotheses he found himself compelled to make, led him afterwards to withdraw it.

The same spirit is found throughout his whole work. He throws into relief the essential — *i.e.*, what is common to all theories; everything that suits only a particular theory is passed over almost in silence. The reader therefore finds himself in the presence of form nearly devoid of matter, which at first he is tempted to take as a fugitive and unassailable phantom. But the efforts he is thus compelled to make force him to think, and eventually he sees that there is often something rather artificial in the theoretical   "aggregates"   which   he   once   admired.

Footnotes

1. This chapter is mainly taken from the prefaces of two of my books — *Théorie Mathématique de la lumiére* (Paris: Naud, 1889), and *Électricité et Optique* (Paris: Naud, 1901).

2. We may add that U will depend only on the q parameters, that T will depend on them and their derivatives with respect to time, and will be a homogeneous polynomial of the second degree with respect to these derivatives.

CHAPTER XIII.
ELECTRO-DYNAMICS.

THE history of electro-dynamics is very instructive from our point of view. The title of Ampère's immortal work is, *Théorie des phéenomènes electro-dynamiques, uniquement fondée sur expérience*. He therefore imagined that he had made no hypotheses; but as we shall not be long in recognising, he was mistaken; only, of these hypotheses he was quite unaware. On the other hand, his successors see them clearly enough, because their attention is attracted by the weak points in Ampère's solution. They made fresh hypotheses, but this time deliberately. How many times they had to change them before they reached the classic system, which is perhaps even now not quite definitive, we shall see.

I. *Ampère's Theory*. — In Ampère's experimental study of the mutual action of currents, he has operated, and he could operate only, with closed currents. This was not because he denied the existence or possibility of open currents. If two conductors are positively and negatively charged and brought into communication by a wire, a current is set up which passes from one to the other until the two potentials are equal. According to the ideas of Ampère's time, this was considered to be an open current; the current was known to pass from the first conductor to the second, but they did not know it returned from the second to the first. All currents of this kind were therefore considered by Ampère to be open currents — for instance, the currents of discharge of a condenser; he was unable to experiment on them, their duration being too short. Another kind of open current may be imagined. Suppose we have two conductors A and B connected by a wire AMB. Small conducting masses in motion are first of all placed in contact with the conductor B, receive an electric charge, and leaving B are set in motion along a path BNA, carrying their charge with them. On coming into contact with A they lose their charge, which then returns to B along the wire AMB. Now here we have, in a sense, a closed circuit, since the electricity describes the closed circuit BNAMB; but the two parts of the current are quite different. In the wire AMB the electricity is displaced *through* a fixed conductor like a voltaic current, overcoming an ohmic resistance and developing heat; we say that it is displaced by *conduction*. In the part BNA the electricity is *carried* by a moving conductor, and is said to be displaced by *convection*. If therefore the convection current is considered to be perfectly analogous to the conduction current, the circuit BNAMB is closed; if on the contrary the convection current is not a "true current," and, for instance, does not act on the magnet, there is only the conduction current AMB, which is *open*. For example, if we connect by a wire the poles of a Holtz machine, the charged rotating disc transfers the electricity by convection from one pole to the other, and it returns to the first pole by conduction through the wire. But currents of this kind are very difficult to produce with appreciable intensity; in fact, with the means at Ampère's disposal we may almost say it was impossible.

To sum up, Ampère could conceive of the existence of two kinds of open currents, but he could experiment on neither, because they were not strong enough, or because their duration was too short. Experiment therefore could only show him the action of a closed current on a closed current — or more accurately, the action of a *closed* current on a portion of current, because a current can be made to describe a closed circuit, of which part may be in motion and the other part fixed. The displacements of the moving part may be studied under the action of another closed current. On the other hand, Ampère had no means of studying the action of an open current either on a closed or on another open current.

1. *The Case of Closed Currents.* — In the case of the mutual action of two closed currents, experiment revealed to Ampère remarkably simple laws. The following will be useful to us in the sequel: — (1) *If the intensity of the currents is kept constant,* and if the two circuits, after having undergone any displacements and deformations whatever, return finally to their initial positions, the total work done by the electro-dynamical actions is zero. In other words, there is an *electro-dynamical potential* of the two circuits proportional to the product of their intensities, and depending on the form and relative positions of the circuits; the work done by the electro-dynamical actions is equal to the change of this potential.

(2) The action of a closed solenoid is zero.

(3) The action of a circuit C on another voltaic circuit C' depends only on the "magnetic field" developed by the circuit C. At each point in space we can, in fact, define in magnitude and direction a certain force called "magnetic force," which enjoys the following properties: —

(*a*) The force exercised by C on a magnetic pole is applied to that pole, and is equal to the magnetic force multiplied by the magnetic mass of the pole.

(*b*) A very short magnetic needle tends to take the direction of the magnetic force, and the couple to which it tends to reduce is proportional to the product of the magnetic force, the magnetic moment of the needle, and the sine of the dip of the needle.

(*c*) If the circuit C is displaced, the amount of the work done by the electro-dynamic action of C on C' will be equal to the increment of "flow of magnetic force" which passes through the circuit.

2. *Action of a Closed Current on a Portion of Current.* — Ampère being unable to produce the open current properly so called, had only one way of studying the action of a closed current on a portion of current. This was by operating on a circuit C composed of two parts, one movable and the other fixed. The movable part was, for instance, a movable wire αβ, the ends α and β of which could slide along a fixed wire. In one of the positions of the movable wire the end α rested on the point A, and the end β on the point B of the fixed wire. The current ran from α to β — *i.e.,* from A to B along the movable wire, and then from B to A along the fixed wire. *This current was therefore closed.*

In the second position, the movable wire having slipped, the points α and β were respectively at A' and B' on the fixed wire. The current ran from α to β — *i.e.,* from A' to B' on the movable wire, and returned from B' to B, and then from B to A, and then from A to A all on the fixed wire. This current was also closed. If a similar circuit be exposed to the action of a closed current C, the movable part will be dis placed just as if it were acted on by a force. Ampère admits that the force, apparently acting on the movable part A B, representing the action of C on the portion αβ of the current, remains the same whether an open current runs through αβ, stopping at α and β, or whether a closed current runs first to β and then returns to α through the fixed portion of the circuit. This hypothesis seemed natural enough, and Ampère innocently assumed it; nevertheless the hypothesis *is not a necessity,* for we shall presently see that Helmholtz rejected it. However that may be, it enabled Ampère, although he had never produced an open current, to lay down the laws of the action of a closed current on an open current, or even on an element of current. They are simple:

(1) The force acting on an element of current is applied to that element; it is normal to the element and to the magnetic force, and proportional to that component of the magnetic force which is normal to the element.

(2) The action of a closed solenoid on an element of current is zero. But the electro-dynamic potential has disappeared — *i.e.*, when a closed and an open current of constant intensities return to their initial positions, the total work done is not zero.

3. *Continuous Rotations.* — The most remarkable electro-dynamical experiments are those in which continuous rotations are produced, and which are called *unipolar induction* experiments. A magnet may turn about its axis; a current passes first through a fixed wire and then enters the magnet by the pole N, for instance, passes through half the magnet, and emerges by a sliding con tact and re-enters the fixed wire. The magnet then begins to rotate continuously. This is Faraday's experiment. How is it possible? If it were a question of two circuits of invariable form, C fixed and C' movable about an axis, the latter would never take up a position of continuous rotation; in fact, there is an electro-dynamical potential; there must therefore be a position of equilibrium when the potential is a maximum. Continuous rotations are therefore possible only when the circuit C' is composed of two parts — one fixed, and the other movable about an axis, as in the case of Faraday's experiment. Here again it is convenient to draw a distinction. The passage from the fixed to the movable part, or *vice versa*, may take place either by simple contact, the same point of the movable part remaining constantly in contact with the same point of the fixed part, or by sliding contact, the same point of the movable part coming successively into con tact with the different points of the fixed part.

It is only in the second case that there can be continuous rotation. This is what then happens: — the system tends to take up a position of equilibrium; but, when at the point of reaching that position, the sliding contact puts the moving part in contact with a fresh point in the fixed part; it changes the connexions and therefore the conditions of equilibrium, so that as the position of equilibrium is ever eluding, so to speak, the system which is trying to reach it, rotation may take place indefinitely. Ampère admits that the action of the circuit on the movable part of C' is the same as if the fixed part of C' did not exist, and therefore as if the current passing through the movable part were an open current. He concluded that the action of a closed on an open current, or *vice versa*, that of an open current on a fixed current, may give rise to continuous rotation. But this conclusion depends on the hypothesis which I have enunciated, and to which, as I said above, Helmholtz declined to subscribe.

4. *Mutual Action of Two Open Currents.* — As far as the mutual action of two open currents, and in particular that of two elements of current, is concerned, all experiment breaks down. Ampère falls back on hypothesis. He assumes: (1) that the mutual action of two elements reduces to a force acting along their *join*; (2) that the action of two closed currents is the resultant of the mutual actions of their different elements, which are the same as if these elements were isolated.

The remarkable thing is that here again Ampère makes two hypotheses without being aware of it. However that may be, these two hypotheses, together with the experiments on closed currents, suffice to determine completely the law of mutual action of two elements. But then, most of the simple laws we have met in the case of closed currents are no longer true. In the first place, there is no electro-dynamical potential; nor was there any, as we have seen, in the case of a closed current acting on an open current. Next, there is, properly speaking, no magnetic force; and we have above denned this force in three different ways: (1) By the action on a magnetic pole; (2) by the director couple which orientates the magnetic needle; (3) by the action on an element of current.

In the case with which we are immediately concerned, not only are these three definitions not in harmony, but each has lost its meaning: —

(1) A magnetic pole is no longer acted on by a unique force applied to that pole. We have seen, in fact, the action of an element of current on a pole is not applied to the pole but to the element; it may, moreover, be replaced by a force applied to the pole and by a couple.

(2) The couple which acts on the magnetic needle is no longer a simple director couple, for its moment with respect to the axis of the needle is not zero. It decomposes into a director couple, properly so called, and a supplementary couple which tends to produce the continuous rotation of which we have spoken above.

(3) Finally, the force acting on an element of a current is not normal to that element. In other words, *the unity of the magnetic force has disappeared.*

Let us see in what this unity consists. Two systems which exercise the same action on a magnetic pole will also exercise the same action on an indefinitely small magnetic needle, or on an element of current placed at the point in space at which the pole is. Well, this is true if the two systems only contain closed currents, and according to Ampère it would not be true if the systems contained open currents. It is sufficient to remark, for instance, that if a magnetic pole is placed at A and an element at B, the direction of the element being in AB produced, this element, which will exercise no action on the pole, will exercise an action either on a magnetic needle placed at A, or on an element of current at A.

5. *Induction.* — We know that the discovery of electro-dynamical induction followed not long after the immortal work of Ampère. As long as it is only a question of closed currents there is no difficulty, and Helmholtz has even remarked that the principle of the conservation of energy is sufficient for us to deduce the laws of induction from the electro-dynamical laws of Ampère. But on the condition, as Bertrand has shown, — that we make a certain number of hypotheses.

The same principle again enables this deduction to be made in the case of open currents, although the result cannot be tested by experiment, since such currents cannot be produced.

If we wish to compare this method of analysis with Ampère's theorem on open currents, we get results which are calculated to surprise us. In the first place, induction cannot be deduced from the variation of the magnetic field by the well-known formula of scientists and practical men; in fact, as I have said, properly speaking, there is no magnetic field. But further, if a circuit C is subjected to the induction of a variable voltaic system S, and if this system S be displaced and deformed in any way whatever, so that the intensity of the currents of this system varies according to any law whatever, then so long as after these variations the system eventually returns to its initial position, it seems natural to suppose that the *mean* electro-motive force induced in the current C is zero. This is true if the circuit C is closed, and if the system S only contains closed currents. It is no longer true if we accept the theory of Ampère, since there would be open currents. So that not only will induction no longer be the variation of the flow of magnetic force in any of the usual senses of the word, but it cannot be represented by the variation of that force whatever it may be.

II. *Helmholtz's Theory.* — I have dwelt upon the consequences of Ampère's theory and on his method of explaining the action of open currents. It is difficult to disregard the paradoxical and artificial character of the propositions to which we are thus led. We feel bound to think "it cannot be so." We may imagine then that Helmholtz has been led to look for something else. He rejects the fundamental hypothesis of Ampère — namely, that the mutual action of two elements of current reduces to a force along their join. He admits that an element of current is not acted upon by a single force but by a force and a couple, and this is what gave rise to the celebrated polemic between Bertrand and Helmholtz. Helmholtz replaces Ampère's hypothesis by the following: — Two elements of current always admit of an electro-dynamic potential, depending solely upon their position and orientation; and the work of the forces that they exercise one on the other is equal to the variation of this potential. Thus Helmholtz can no more do without hypothesis than Ampère, but at least he does not do so without explicitly announcing it. In the case of closed currents, which alone are accessible to experiment, the two theories agree; in all other cases they differ. In the first place, contrary to what Ampère supposed, the force which seems to act on the movable portion of a closed current is not the same as that acting on the movable portion if it were isolated and if it constituted an open current. Let us return to the circuit C', of which we spoke above, and which was formed of a movable wire sliding on a fixed wire. In the only experiment that can be made the movable portion αβ is not isolated, but is part of a closed circuit. When it passes from AB to A'B', the total electro-dynamic potential varies for two reasons. First, it has a slight increment because the potential of A'B' with respect to the circuit C is not the same as that of AB; secondly, it has a second increment because it must be increased by the potentials of the elements AA' and B'B with respect to C. It is this *double* increment which represents the work of the force acting upon the portion AB. If, on the contrary, αβ be isolated, the potential would only have the first increment, and this first increment alone would measure the work of the force acting on AB. In the second place, there could be no continuous rotation without sliding contact, and in fact, that, as we have seen in the case of closed currents, is an immediate consequence of the existence of an electro-dynamic potential. In Faraday's experiment, if the magnet is fixed, and if the part of the current external to the magnet runs along a movable wire, that movable wire may undergo continuous rotation. But it does not mean that, if the contacts of the weir with the magnet were suppressed, and an open current were to run along the wire, the wire would still have a movement of continuous rotation. I have just said, in fact, that an isolated element is not acted on in the same way as a movable element making part of a closed circuit. But there is another difference. The action of a solenoid on a closed current is zero according to experiment and according to the two theories. Its action on an open current would be zero according to Ampère, and it would not be zero according to Helmholtz. From this follows an important consequence. We have given above three definitions of the magnetic force. The third has no meaning here, since an element of current is no longer acted upon by a single force. Nor has the first any meaning. What, in fact, is a magnetic pole? It is the extremity of an indefinite linear magnet. This magnet may be replaced by an indefinite solenoid. For the definition of magnetic force to have any meaning, the action exercised by an open current on an indefinite solenoid would only depend on the position of the extremity of that solenoid — *i.e.*, that the action of a closed solenoid is zero. Now we have just seen that this is not the case. On the other hand, there is nothing to prevent us from adopting the second definition which is founded on the measurement of the director couple which tends to orientate the magnetic needle; but, if it is adopted, neither the effects of induction nor electro-dynamic effects will depend solely on the distribution of the lines of force in this magnetic field.

III. *Difficulties raised by these Theories.* — Helmholtz's theory is an advance on that of Ampère; it is necessary, however, that every difficulty should be removed. In both, the name "magnetic field" has no meaning, or, if we give it one by a more or less artificial convention, the ordinary laws so familiar to electricians no longer apply; and it is thus that the electro-motive force induced in a wire is no longer measured by the number of lines of force met by that wire. And our objections do not proceed only from the fact that it is difficult to give up deeply-rooted habits of language and thought. There is something more. If we do not believe in actions at a distance, electro-dynamic phenomena must be explained by a modification of the medium. And this medium is precisely what we call "magnetic field," and then the electro-magnetic effects should only depend on that field. All these difficulties arise from the hypothesis of open currents.

IV. *Maxwell's Theory.* — Such were the difficulties raised by the current theories, when Maxwell with a stroke of the pen caused them to vanish. To his mind, in fact, all currents are closed currents. Maxwell admits that if in a dielectric, the electric field happens to vary, this dielectric becomes the seat of a particular phenomenon acting on the galvanometer like a current and called the *current of displacement*. If, then, two conductors bearing positive and negative charges are placed in connection by means of a wire, during the discharge there is an open current of conduction in that wire; but there are produced at the same time in the surrounding dielectric currents of displacement which close this current of conduction. We know that Maxwell's theory leads to the explanation of optical phenomena which would be due to extremely rapid electrical oscillations. At that period such a conception was only a daring hypothesis which could be supported by no experiment; but after twenty years Maxwell's ideas received the confirmation of experiment. Hertz succeeded in producing systems of electric oscillations which reproduce all the properties of light, and only differ by the length of their wave — that is to say, as violet differs from red. In some measure he made a synthesis of light. It might be said that Hertz has not directly proved Maxwell's fundamental idea of the action of the current of displacement on the galvanometer. That is true in a sense. What he has shown directly is that electro-magnetic induction is not instantaneously propagated, as was supposed, but its speed is the speed of light. Yet, to suppose there is no current of displacement, and that induction is with the speed of light; or, rather, to suppose that the currents of displacement produce inductive effects, and that the induction takes place instantaneously — *comes to the same thing*. This cannot be seen at the first glance, but it is proved by an analysis of which I must not even think of giving even a summary here.

V. *Rowland's Experiment*. — But, as I have said above, there are two kinds of open conduction currents. There are first the currents of discharge of a condenser, or of any conductor whatever. There are also cases in which the electric charges describe a closed contour, being displaced by conduction in one part of the circuit and by convection in the other part. The question might be regarded as solved for open currents of the first kind; they were closed by currents of displacement. For open currents of the second kind the solution appeared still more simple. It seemed that if the current were closed it could only be by the current of convection itself. For that purpose it was sufficient to admit that a "convection current" — *i.e.*, a charged conductor in motion could act on the galvanometer. But experimental confirmation was lacking. It appeared difficult, in fact, to obtain a sufficient intensity even by increasing as much as possible the charge and the velocity of the conductors. Rowland, an extremely skilful experimentalist, was the first to triumph, or to seem to triumph, over these difficulties. A disc received a strong electrostatic charge and a very high speed of rotation. An astatic magnetic system placed beside the disc underwent deviations. The experiment was made twice by Rowland, once in Berlin and once at Baltimore. It was afterwards repeated by Himstedt. These physicists even believed that they could announce that they had succeeded in making quantitative measurements. For twenty years Rowland's law was admitted without objection by all physicists, and, indeed, everything seemed to confirm it. The spark certainly does produce a magnetic effect, and does it not seem extremely likely that the spark discharged is due to particles taken from one of the electrodes and transferred to the other electrode with their charge? Is not the very spectrum of the spark, in which we recognise the lines of the metal of the electrode, a proof of it? The spark would then be a real current of induction. On the other hand, it is also admitted that in an electrolyte the electricity is carried by the ions in motion. The current in an electrolyte would therefore also be a current of convection; but it acts on the magnetic needle. And in the same way for cathodic rays; Crooks attributed these rays to very subtle matter charged with negative electricity and moving with very high velocity. He looked upon them, in other words, as currents of convection. Now, these cathodic rays are deviated by the magnet. In virtue of the principle of action and re-action, they should in their turn deviate the magnetic needle. It is true that Hertz believed he had proved that the cathodic rays do not carry negative electricity, and that they do not act on the magnetic needle; but Hertz was wrong. First of all, Perrin succeeded in collecting the electricity carried by these rays — electricity of which Hertz denied the existence; the German scientist appears to have been deceived by the effects due to the action of the X-rays, which were not yet discovered. Afterwards, and quite recently, the action of the cathodic rays on the magnetic needle has been brought to light. Thus all these phenomena looked upon as currents of convection, electric sparks, electrolytic currents, cathodic rays, act in the same manner on the galvanometer and in conformity to Rowland's law.

VI. *Lorentz's Theory*. We need not go much further. According to Lorentz's theory, currents of conduction would themselves be true convection currents. Electricity would remain indissolubly connected with certain material particles called *electrons*. The circulation of these electrons through bodies would produce voltaic currents, and what would distinguish conductors from insulators would be that the one could be traversed by these electrons, while the others would check the movement of the electrons. Lorentz's theory is very attractive. It gives a very simple explanation of certain phenomena, which the earlier theories — even Maxwell's in its primitive form — could only deal with in an unsatisfactory manner; for example, the aberration of light, the partial impulse of luminous waves, magnetic polarisation, and Zeeman's experiment.

A few objections still remained. The phenomena of an electric system seemed to depend on the absolute velocity of translation of the centre of gravity of this system, which is contrary to the idea that we have of the relativity of space. Supported by M. Crémieu, M. Lippman has presented this objection in a very striking form. Imagine two charged conductors with the same velocity of translation. They are relatively at rest. However, each of them being equivalent to a current of convection, they ought to attract one another, and by measuring this attraction we could measure their absolute velocity. "No!" replied the partisans of Lorentz. "What we could measure in that way is not their absolute velocity, but their relative velocity *with respect to the ether*, so that the principle of relativity is safe." Whatever there may be in these objections, the edifice of electro-dynamics seemed, at any rate in its broad lines, definitively constructed. Everything was presented under the most satisfactory aspect. The theories of Ampère and Helmholtz, which were made for the open currents that no longer existed, seem to have no more than purely historic interest, and the in extricable complications to which these theories led have been almost forgotten. This quiescence has been recently disturbed by the experiments of M. Crémieu, which have contradicted, or at least have seemed to contradict, the results formerly obtained by Rowland. Numerous investigators have endeavoured to solve the question, and fresh experiments have been undertaken. What result will they give? I shall take care not to risk a prophecy which might be falsified between the day this book is ready for the press and the day on which it is placed before the public.

THE END.

The Principles of Mathematical Physics (1904)
by Henri Poincaré, translated by George Bruce Halsted

The History of Mathematical Physics

The Past and the Future of Physics.

What is the present state of mathematical physics? What are the problems it is led to set itself? What is its future? Is its orientation about to be modified?

Ten years hence will the aim and the methods of this science appear to our immediate successors in the same light as to ourselves; or, on the contrary, are we about to witness a profound transformation? Such are the questions we are forced to raise in entering to-day upon our investigation.

If it is easy to propound them: to answer is difficult. If we felt tempted to risk a prediction, we should easily resist this temptation, by thinking of all the stupidities the most eminent savants of a hundred years ago would have uttered, if some one had asked them what the science of the nineteenth century would be. They would have thought themselves bold in their predictions, and after the event, how very timid we should have found them. Do not, therefore, expect of me any prophecy.

But if, like all prudent physicians, I shun giving a prognosis, yet I can not dispense with a little diagnostic; well, yes, there are indications of a serious crisis, as if we might expect an approaching transformation. Still, be not too anxious: we are sure the patient will not die of it, and we may even hope that this crisis will be salutary, for the history of the past seems to guarantee us this. This crisis, in fact, is not the first, and to understand it, it is important to recall those which have preceded. Pardon then a brief historical sketch.

The Physics of Central Forces.

Mathematical physics, as we know, was born of celestial mechanics, which gave birth to it at the end of the eighteenth century, at the moment when it itself attained its complete development. During its first years especially, the infant strikingly resembled its mother.

The astronomic universe is formed of masses, very great, no doubt, but separated by intervals so immense that they appear to us only as material points. These points attract each other inversely as the square of the distance, and this attraction is the sole force which influences their movements. But if our senses were sufficiently keen to show us all the details of the bodies which the physicist studies, the spectacle thus disclosed would scarcely differ from the one the astronomer contemplates. There also we should see material points, separated from one another by intervals, enormous in comparison with their dimensions, and describing orbits according to regular laws. These infinitesimal stars are the atoms. Like the stars proper, they attract or repel each other, and this attraction or this repulsion, following the straight line which joins them, depends only on the distance. The law according to which this force varies as function of the distance is perhaps not the law of Newton, but it is an analogous law; in place of the exponent —2, we have probably a different exponent, and it is from this change of exponent that arises all the diversity of physical phenomena, the variety of qualities and of sensations, all the world, colored and sonorous, which surrounds us; in a word, all nature.

Such is the primitive conception in all its purity. It only remains to seek in the different cases what value should be given to this exponent in order to explain all the facts. It is on this model that Laplace, for example, constructed his beautiful theory of capillarity; he regards it only as a particular case of attraction, or, as he says, of universal gravitation, and no one is astonished to find it in the middle of one of the five volumes of the 'Mécanique céleste.' More recently Briot believes he penetrated the final secret of optics in demonstrating that the atoms of ether attract each other in the inverse ratio of the sixth power of the distance; and Maxwell himself, does he not say somewhere that the atoms of gases repel each other in the inverse ratio of the fifth power of the distance? We have the exponent —6, or —5, in place of the exponent —2, but it is always an exponent.

Among the theories of this epoch, one alone is an exception, that of Fourier; in it are indeed atoms acting at a distance one upon the other; they mutually transmit heat, but they do not attract, they never budge. From this point of view, Fourier's theory must have appeared to the eyes of his contemporaries, to those of Fourier himself, as imperfect and provisional.

This conception was not without grandeur; it was seductive, and many among us have not finally renounced it; they know that one will attain the ultimate elements of things only by patiently disentangling the complicated skein that our senses give us; that it is necessary to advance step by step, neglecting no intermediary; that our fathers were wrong in wishing to skip stations; but they believe that when one shall have arrived at these ultimate elements, there again will be found the majestic simplicity of celestial mechanics.

Neither has this conception been useless; it has rendered us an inestimable service, since it has contributed to make precise the fundamental notion of the physical law.

I will explain myself; how did the ancients understand law? It was for them an internal harmony, static, so to say, and immutable; or else it was like a model that nature tried to imitate. For us a law is something quite different; it is a constant relation between the phenomenon of to-day and that of to-morrow; in a word, it is a differential equation.

Behold the ideal form of physical law; well, it is Newton's law which first clothed it forth. If then one has acclimated this form in physics, it is precisely by copying as far as possible this law of Newton, that is by imitating celestial mechanics. This is, moreover, the idea I have tried to bring out in Chapter VI.

The Physics of the Principles.

Nevertheless, a day arrived when the conception of central forces no longer appeared sufficient, and this is the first of those crises of which I just now spoke.

What was done then? The attempt to penetrate into the detail of the structure of the universe, to isolate the pieces of this vast mechanism, to analyze one by one the forces which put them in motion, was abandoned, and we were content to take as guides certain general principles the express object of which is to spare us this minute study. How so? Suppose we have before us any machine; the initial wheel work and the final wheel work alone are visible, but the transmission, the intermediary machinery by which the movement is communicated from one to the other, is hidden in the interior and escapes our view; we do not know whether the communication is made by gearing or by belts, by connecting-rods or by other contrivances. Do we say that it is impossible for us to understand anything about this machine so long as we are not permitted to take it to pieces? You know well we do not, and that the principle of the conservation of energy suffices to determine for us the most interesting point. We easily ascertain that the final wheel turns ten times less quickly than the initial wheel, since these two wheels are visible; we are able thence to conclude that a couple applied to the one will be balanced by a couple ten times greater applied to the other. For that there is no need to penetrate the mechanism of this equilibrium and to know how the forces compensate each other in the interior of the machine; it suffices to be assured that this compensation can not fail to occur.

Well, in regard to the universe, the principle of the conservation of energy is able to render us the same service. The universe is also a machine, much more complicated than all those of industry, of which almost all the parts are profoundly hidden from us; but in observing the motion of those that we can see, we are able, by the aid of this principle, to draw conclusions which remain true whatever may be the details of the invisible mechanism which animates them.

The principle of the conservation of energy, or Mayer's principle, is certainly the most important, but it is not the only one; there are others from which we can derive the same advantage. These are:

Carnot's principle, or the principle of the degradation of energy.

Newton's principle, or the principle of the equality of action and reaction.

The principle of relativity, according to which the laws of physical phenomena must be the same for a stationary observer as for an observer carried along in a uniform motion of translation; so that we have not and can not have any means of discerning whether or not we are carried along in such a motion.

The principle of the conservation of mass, or Lavoisier's principle.

I will add the principle of least action.

The application of these five or six general principles to the different physical phenomena is sufficient for our learning of them all that we could reasonably hope to know of them. The most remarkable example of this new mathematical physics is, beyond question, Maxwell's electromagnetic theory of light.

We know nothing as to what the ether is, how its molecules are disposed, whether they attract or repel each other; but we know that this medium transmits at the same time the optical perturbations and the electrical perturbations; we know that this transmission must take place in conformity with the general principles of mechanics, and that suffices us for the establishment of the equations of the electromagnetic field.

These principles are results of experiments boldly generalized; but they seem to derive from their very generality a high degree of certainty. In fact, the more general they are, the more frequent are the opportunities to check them, and the verifications multiplying, taking the most varied, the most unexpected forms, end by no longer leaving place for doubt.

Utility of the Old Physics.

Such is the second phase of the history of mathematical physics and we have not yet emerged from it. Shall we say that the first has been useless? that during fifty years science went the wrong way, and that there is nothing left but to forget so many accumulated efforts that a vicious conception condemned in advance to failure? Not the least in the world. Do you think the second phase could have come into existence without the first? The hypothesis of central forces contained all the principles; it involved them as necessary consequences; it involved both the conservation of energy and that of masses, and the equality of action and reaction, and the law of least action, which appeared, it is true, not as experimental truths, but as theorems; the enunciation of which had at the same time something more precise and less general than under their present form.

It is the mathematical physics of our fathers which has familiarized us little by little with these various principles; which has habituated us to recognize them under the different vestments in which they disguise themselves. They have been compared with the data of experience, it has been seen how it was necessary to modify their enunciation to adapt them to these data; thereby they have been extended and consolidated. Thus they came to be regarded as experimental truths; the conception of central forces became then a useless, support, or rather an embarrassment, since it made the principles partake of its hypothetical character.

The frames then have not broken, because they are elastic; but they have enlarged; our fathers, who established them, did not labor in vain, and we recognize in the science of to-day the general traits of the sketch which they traced.

The Present Crisis of Mathematical Physics

The New Crisis.

Are we now about to enter upon a third period? Are we on the eve of a second crisis? These principles on which we have built all, are they about to crumble away in their turn? This has been for some time a pertinent question.

When I speak thus, you no doubt think of radium, that grand revolutionist of the present time, and in fact I shall come back to it presently; but there is something else. It is not alone the conservation of energy which is in question; all the other principles are equally in danger, as we shall see in passing them successively in review.

Carnot's Principle. — Let us commence with the principle of Carnot. This is the only one which does not present itself as an immediate consequence of the hypothesis of central forces; more than that, it seems, if not to directly contradict that hypothesis, at least not to be reconciled with it without a certain effort. If physical phenomena were due exclusively to the movements of atoms whose mutual attraction depended only on the distance, it seems that all these phenomena should be reversible; if all the initial velocities were reversed, these atoms, always subjected to the same forces, ought to go over their trajectories in the contrary sense, just as the earth would describe in the retrograde sense this same elliptic orbit which it describes in the direct sense, if the initial conditions of its motion had been reversed. On this account, if a physical phenomenon is possible, the inverse phenomenon should be equally so, and one should be able to reascend the course of time. Now, it is not so in nature, and this is precisely what the principle of Carnot teaches us; heat can pass from the warm body to the cold body; it is impossible afterward to make it take the inverse route and to reestablish differences of temperature which have been effaced. Motion can be wholly dissipated and transformed into heat by friction; the contrary transformation can never be made except partially.

We have striven to reconcile this apparent contradiction. If the world tends toward uniformity, this is not because its ultimate parts, at first unlike, tend to become less and less different; it is because, shifting at random, they end by blending. For an eye which should distinguish all the elements, the variety would remain always as great; each grain of this dust preserves its originality and does not model itself on its neighbors; but as the blend becomes more and more intimate, our gross senses perceive only the uniformity. This is why for example, temperatures tend to a level, without the possibility of going backwards.

A drop of wine falls into a glass of water; whatever may be the law of the internal motion of the liquid, we shall soon see it colored of a uniform rosy tint, and however much from this moment one may shake it afterwards, the wine and the water do not seem capable of again separating. Here we have the type of the irreversible physical phenomenon: to hide a grain of barley in a heap of wheat, this is easy; afterwards to find it again and get it out, this is practically impossible. All this Maxwell and Boltzmann have explained; but the one who has seen it most clearly, in a book too little read because it is a little difficult to read, is Gibbs, in his 'Elementary Principles of Statistical Mechanics.'

For those who take this point of view, Carnot's principle is only an imperfect principle, a sort of concession to the infirmity of our senses; it is because our eyes are too gross that we do not distinguish the elements of the blend; it is because our hands are too gross that we can not force them to separate; the imaginary demon of Maxwell, who is able to sort the molecules one by one, could well constrain the world to return backward. Can it return of itself? That is not impossible; that is only infinitely improbable. The chances are that we should wait a long time for the concourse of circumstances which would permit a retrogradation; but sooner or later they will occur, after years whose number it would take millions of figures to write. These reservations, however, all remained theoretic; they were not very disquieting, and Carnot's principle retained all its practical value. But here the scene changes. The biologist, armed with his microscope, long ago noticed in his preparations irregular movements of little particles in suspension; this is the Brownian movement. He first thought this was a vital phenomenon, but soon he saw that the inanimate bodies danced with no less ardor than the others; then he turned the matter over to the physicists. Unhappily, the physicists remained long uninterested in this question; one concentrates the light to illuminate the microscopic preparation, thought they; with light goes heat; thence inequalities of temperature and in the liquid interior currents which produce the movements referred to. It occurred to M. Gouy to look more closely, and he saw, or thought he saw, that this explanation is untenable, that the movements become brisker as the particles are smaller, but that they are not influenced by the mode of illumination. If then these movements never cease, or rather are reborn without cease, without borrowing anything from an external source of energy, what ought we to believe? To be sure, we should not on this account renounce our belief in the conservation of energy, but we see under our eyes now motion transformed into heat by friction, now inversely heat changed into motion, and that without loss since the movement lasts forever. This is the contrary of Carnot's principle. If this be so, to see the world return backward, we no longer have need of the infinitely keen eye of Maxwell's demon; our microscope suffices. Bodies too large, those, for example, which are a tenth of a millimeter, are hit from all sides by moving atoms, but they do not budge, because these shocks are very numerous and the law of chance makes them compensate each other; but the smaller particles receive too few shocks for this compensation to take place with certainty and are incessantly knocked about. And behold already one of our principles in peril.

The Principle of Relativity.

Let us pass to the principle of relativity: this not only is confirmed by daily experience, not only is it a necessary consequence of the hypothesis of central forces, but it is irresistibly imposed upon our good sense, and yet it also is assailed. Consider two electrified bodies; though they seem to us at rest, they are both carried along by the motion of the earth; an electric charge in motion, Rowland has taught us, is equivalent to a current; these two charged bodies are, therefore, equivalent to two parallel currents of the samesense and these two currents should attract each other. In measuring this attraction, we shall measure the velocity of the earth; not its velocity in relation to the sun or the fixed stars, but its absolute velocity.

I well know what will be said: It is not its absolute velocity that is measured, it is its velocity in relation to the ether. How unsatisfactory that is! Is it not evident that from the principle so understood we could no longer infer anything? It could no longer tell us anything just because it would no longer fear any contradiction. If we succeed in measuring anything, we shall always be free to say that this is not the absolute velocity, and if it is not the velocity in relation to the ether, it might always be the velocity in relation to some new unknown fluid with which we might fill space.

Indeed, experiment has taken upon itself to ruin this interpretation of the principle of relativity; all attempts to measure the velocity of the earth in relation to the ether have led to negative results. This time experimental physics has been more faithful to the principle than mathematical physics; the theorists, to put in accord their other general views, would not have spared it; but experiment has been stubborn in confirming it. The means have been varied; finally Michelson pushed precision to its last limits; nothing came of it. It is precisely to explain this obstinacy that the mathematicians are forced to-day to employ all their ingenuity.

Their task was not easy, and if Lorentz has got through it, it is only by accumulating hypotheses.

The most ingenious idea was that of local time. Imagine two observers who wish to adjust their timepieces by optical signals; they exchange signals, but as they know that the transmission of light is not instantaneous, they are careful to cross them. When station B perceives the signal from station A, its clock should not mark the same hour as that of station A at the moment of sending the signal, but this hour augmented by a constant representing the duration of the transmission. Suppose, for example, that station A sends its signal when its clock marks the hour *0*, and that station B perceives it when its clock marks the hour *t*. The clocks are adjusted if the slowness equal to *t* represents the duration of the transmission, and to verify it, station B sends in its turn a signal when its clock marks 0; then station A should perceive it when its clock marks *t*. The timepieces are then adjusted.

And in fact they mark the same hour at the same physical instant, but on the one condition, that the two stations are fixed. Otherwise the duration of the transmission will not be the same in the two senses, since the station A, for example, moves forward to meet the optical perturbation emanating from B, whereas the station B flees before the perturbation emanating from A. The watches adjusted in that way will not mark, therefore, the true time; they will mark what may be called the *local time*, so that one of them will be slow of the other. It matters little, since we have no means of perceiving it. All the phenomena which happen at A, for example, will be late, but all will be equally so, and the observer will not perceive it, since his watch is slow; so, as the principle of relativity requires, he will have no means of knowing whether he is at rest or in absolute motion.

Unhappily, that does not suffice, and complementary hypotheses are necessary; it is necessary to admit that bodies in motion undergo a uniform contraction in the sense of the motion. One of the diameters of the earth, for example, is shrunk by one two-hundred-millionth in consequence of our planet's motion, while the other diameter retains its normal length. Thus the last little differences are compensated. And then, there is still the hypothesis about forces. Forces, whatever be their origin, gravity as well as elasticity, would be reduced in a certain proportion in a world animated by a uniform translation; or, rather, this would happen for the components perpendicular to the translation; the components parallel would not change. Resume, then, our example of two electrified bodies; these bodies repel each other, but at the same time if all is carried along in a uniform translation, they are equivalent to two parallel currents of the same sense which attract each other. This electrodynamic attraction diminishes, therefore, the electrostatic repulsion, and the total repulsion is feebler than if the two bodies were at rest. But since to measure this repulsion we must balance it by another force, and all these other forces are reduced in the same proportion, we perceive nothing. Thus all seems arranged, but are all the doubts dissipated? What would happen if one could communicate by non-luminous signals whose velocity of propagation differed from that of light? If, after having adjusted the watches by the optical procedure, we wished to verify the adjustment by the aid of these new signals, we should observe discrepancies which would render evident the common translation of the two stations. And are such signals inconceivable, if we admit with Laplace that universal gravitation is transmitted a million times more rapidly than light?

Thus, the principle of relativity has been valiantly defended in these latter times, but the very energy of the defense proves how serious was the attack.

Newton's Principle.

Let us speak now of the principle of Newton, on the equality of action and reaction. This is intimately bound up with the preceding, and it seems indeed that the fall of the one would involve that of the other. Thus we must not be astonished to find here the same difficulties.

Electrical phenomena, according to the theory of Lorentz, are due to the displacements of little charged particles, called electrons, immersed in the medium we call ether. The movements of these electrons produce perturbations in the neighboring ether; these perturbations propagate themselves in every direction with the velocity of light, and in turn other electrons, originally at rest, are made to vibrate when the perturbation reaches the parts of the ether which touch them. The electrons, therefore, act on one another, but this action is not direct, it is accomplished through the ether as intermediary. Under these conditions can there be compensation between action and reaction, at least for an observer who should take account only of the movements of matter, that is, of the electrons, and who should be ignorant of those of the ether that he could not see? Evidently not. Even if the compensation should be exact, it could not be simultaneous. The perturbation is propagated with a finite velocity; it, therefore, reaches the second electron only when the first has long ago entered upon its rest. This second electron, therefore, will undergo, after a delay, the action of the first, but will certainly not at that moment react upon it, since around this first electron nothing any longer budges.

The analysis of the facts permits us to be still more precise. Imagine, for example, a Hertzian oscillator, like those used in wireless telegraphy; it sends out energy in every direction; but we can provide it with a parabolic mirror, as Hertz did with his smallest oscillators, so as to send all the energy produced in a single direction. What happens then according to the theory? The apparatus recoils, as if it were a cannon and the projected energy a ball; and that is contrary to the principle of Newton, since our projectile here has no mass, it is not matter, it is energy. The case is still the same, moreover, with a beacon light provided with a reflector, since light is nothing but a perturbation of the electromagnetic field. This beacon light should recoil as if the light it sends out were a projectile. What is the force that should produce this recoil? It is what is called the Maxwell-Bartholi pressure. It is very minute, and it has been difficult to put it in evidence even with the most sensitive radiometers; but it suffices that it exists.

If all the energy issuing from our oscillator falls on a receiver, this will act as if it had received a mechanical shock, which will represent in a sense the compensation of the oscillator's recoil; the reaction will be equal to the action, but it will not be simultaneous; the receiver will move on, but not at the moment when the oscillator recoils. If the energy propagates itself indefinitely without encountering a receiver, the compensation will never occur.

Shall we say that the space which separates the oscillator from the receiver and which the perturbation must pass over in going from the one to the other is not void, that it is full not only of ether, but of air, or even in the interplanetary spaces of some fluid subtile but still ponderable; that this matter undergoes the shock like the receiver at the moment when the energy reaches it, and recoils in its turn when the perturbation quits it! That would save Newton's principle, but that is not true. If energy in its diffusion remained always attached to some material substratum, then matter in motion would carry along light with it, and Fizeau has demonstrated that it does nothing of the sort, at least for air. Michelson and Morley have since confirmed this. It might be supposed also that the movements of matter proper are exactly compensated by those of the ether; but that would lead us to the same reflections as before now. The principle so understood will explain everything, since, whatever might be the visible movements, we always could imagine hypothetical movements which compensate them. But if it is able to explain everything, this is because it does not enable us to foresee anything; it does not enable us to decide between the different possible hypotheses, since it explains everything beforehand. It therefore becomes useless.

And then the suppositions that it would be necessary to make on the movements of the ether are not very satisfactory. If the electric charges double, it would be natural to imagine that the velocities of the diverse atoms of ether double also; but, for the compensation, it would be necessary that the mean velocity of the ether quadruple.

This is why I have long thought that these consequences of theory, contrary to Newton's principle, would end some day by being abandoned, and yet the recent experiments on the movements of the electrons issuing from radium seem rather to confirm them.

Lavoisier's Principle.

I arrive at the principle of Lavoisier on the conservation of mass. Certainly, this is one not to be touched without unsettling all mechanics. And now certain persons think that it seems true to us only because in mechanics merely moderate velocities are considered, but that it would cease to be true for bodies animated by velocities comparable to that of light. Now these velocities are believed at present to have been realized; the cathode rays and those of radium may be formed of very minute particles or of electrons which are displaced with velocities smaller no doubt than that of light, but which might be its one tenth or one third.

These rays can be deflected, whether by an electric field, or by a magnetic field, and we are able, by comparing these deflections, to measure at the same time the velocity of the electrons and their mass (or rather the relation of their mass to their charge). But when it was seen that these velocities approached that of light, it was decided that a correction was necessary. These molecules, being electrified, can not be displaced without agitating the ether; to put them in motion it is necessary to overcome a double inertia, that of the molecule itself and that of the ether. The total or apparent mass that one measures is composed, therefore, of two parts: the real or mechanical mass of the molecule and the electrodynamic mass representing the inertia of the ether.

The calculations of Abraham and the experiments of Kaufmann have then shown that the mechanical mass, properly so called, is null, and that the mass of the electrons, or, at least, of the negative electrons, is of exclusively electrodynamic origin. This is what forces us to change the definition of mass; we can not any longer distinguish mechanical mass and electrodynamic mass, since then the first would vanish; there is no mass other than electrodynamic inertia. But in this case the mass can no longer be constant; it augments with the velocity, and it even depends on the direction, and a body animated by a notable velocity will not oppose the same inertia to the forces which tend to deflect it from its route, as to those which tend to accelerate or to retard its progress.

There is still a resource; the ultimate elements of bodies are electrons, some charged negatively, the others charged positively. The negative electrons have no mass, this is understood; but the positive electrons, from the little we know of them, seem much greater. Perhaps they have, besides their electrodynamic mass, a true mechanical mass. The real mass of a body would, then, be the sum of the mechanical masses of its positive electrons, the negative electrons not counting; mass so defined might still be constant.

Alas! this resource also evades us. Recall what we have said of the principle of relativity and of the efforts made to save it. And it is not merely a principle which it is a question of saving, it is the indubitable results of the experiments of Michelson.

Well, as was above seen, Lorentz, to account for these results, was obliged to suppose that all forces, whatever their origin, were reduced in the same proportion in a medium animated by a uniform translation; this is not sufficient; it is not enough that this take place for the real forces, it must also be the same for the forces of inertia; it is therefore necessary, he says, that *the masses of all the particles be influenced by a translation to the same degree as the electromagnetic masses of the electrons.*

So the mechanical masses must vary in accordance with the same laws as the electrodynamic masses; they can not, therefore, be constant.

Need I point out that the fall of Lavoisier's principle involves that of Newton's? This latter signifies that the center of gravity of an isolated system moves in a straight line; but if there is no longer a constant mass, there is no longer a center of gravity, we no longer know even what this is. This is why I said above that the experiments on the cathode rays appeared to justify the doubts of Lorentz concerning Newton's principle.

From all these results, if they were confirmed, would arise an entirely new mechanics, which would be, above all, characterized by this fact, that no velocity could surpass that of light, (¹ Because bodies would oppose an increasing inertia to the causes which would tend to accelerate their motion; and this inertia would become infinite when one approached the velocity of light.) any more than any temperature can fall below absolute zero.

No more for an observer, carried along himself in a translation he does not suspect, could any apparent velocity surpass that of light; and this would be then a contradiction, if we did not recall that this observer would not use the same clocks as a fixed observer, but, indeed, clocks marking 'local time.'

Here we are then facing a question I content myself with stating. If there is no longer any mass, what becomes of Newton's law? Mass has two aspects: it is at the same time a coefficient of inertia and an attracting mass entering as factor into Newtonian attraction. If the coefficient of inertia is not constant, can the attracting mass be? That is the question.

Mayer's Principle.

At least, the principle of the conservation of energy yet remained to us, and this seemed more solid. Shall I recall to you how it was in its turn thrown into discredit? This event has made more noise than the preceding, and it is in all the memoirs. From the first works of Becquerel, and, above all, when the Curies had discovered radium, it was seen that every radioactive body was an inexhaustible source of radiation. Its activity seemed to subsist without alteration throughout the months and the years. This was in itself a strain on the principles; these radiations were in fact energy, and from the same morsel of radium this issued and forever issued. But these quantities of energy were too slight to be measured; at least that was the belief and we were not much disquieted.

The scene changed when Curie bethought himself to put radium in a calorimeter; it was then seen that the quantity of heat incessantly created was very notable.

The explanations proposed were numerous; but in such case we can not say, the more the better. In so far as no one of them has prevailed over the others, we can not be sure there is a good one among them. Since some time, however, one of these explanations seems to be getting the upper hand and we may reasonably hope that we hold the key to the mystery.

Sir W. Ramsay has striven to show that radium is in process of transformation, that it contains a store of energy enormous but not inexhaustible. The transformation of radium then would produce a million times more heat than all known transformations; radium would wear itself out in 1,250 years; this is quite short, and you see that we are at least certain to have this point settled some hundreds of years from now. While waiting, our doubts remain.

The Future of Mathematical Physics

The Principles and Experiment.

In the midst of so much ruin, what remains standing? The principle of least action is hitherto intact, and Larmor appears to believe that it will long survive the others; in reality, it is still more vague and more general.

In presence of this general collapse of the principles, what attitude will mathematical physics take? And first, before too much excitement, it is proper to ask if all that is really true. All these derogations to the principles are encountered only among infinitesimals; the microscope is necessary to see the Brownian movement; electrons are very light; radium is very rare, and one never has more than some milligrams of it at a time. And, then, it may be asked whether, besides the infinitesimal seen, there was not another infinitesimal unseen counterpoise to the first

So there is an interlocutory question, and, as it seems, only experiment can solve it. We shall, therefore, only have to hand over the matter to the experimenters, and, while waiting for them to finally decide the debate, not to preoccupy ourselves with these disquieting problems, and to tranquilly continue our work as if the principles were still uncontested. Certes, we have much to do without leaving the domain where they may be applied in all security; we have enough to employ our activity during this period of doubts.

The Role of the Analyst.

And as to these doubts, is it indeed true that we can do nothing to disembarrass science of them? It must indeed be said, it is not alone experimental physics that has given birth to them; mathematical physics has well contributed. It is the experimenters who have seen radium throw out energy, but it is the theorists who have put in evidence all the difficulties raised by the propagation of light across a medium in motion; but for these it is probable we should not have become conscious of them. Well, then, if they have done their best to put us into this embarrassment, it is proper also that they help us to get out of it

They must subject to critical examination all these new views I have just outlined before you, and abandon the principles only after having made a loyal effort to save them. What can they do in this sense? That is what I will try to explain.

It is a question before all of endeavoring to obtain a more satisfactory theory of the electrodynamics of bodies in motion. It is there especially, as I have sufficiently shown above, that difficulties accumulate. It is useless to heap up hypotheses, we can not satisfy all the principles at once; so far, one has succeeded in safeguarding some only on condition of sacrificing the others; but all hope of obtaining better results is not yet lost. Let us take, then, the theory of Lorentz, turn it in all senses, modify it little by little, and perhaps everything will arrange itself.

Thus in place of supposing that bodies in motion undergo a contraction in the sense of the motion, and that this contraction is the same whatever be the nature of these bodies and the forces to which they are otherwise subjected, could we not make a more simple and natural hypothesis? We might imagine, for example, that it is the ether which is modified when it is in relative motion in reference to the material medium which penetrates it, that, when it is thus modified, it no longer transmits perturbations with the same velocity in every direction. It might transmit more rapidly those which are propagated parallel to the motion of the medium, whether in the same sense or in the opposite sense, and less rapidly those which are propagated perpendicularly. The wave surfaces would no longer be spheres, but ellipsoids, and we could dispense with that extraordinary contraction of all bodies.

I cite this only as an example, since the modifications that might be essayed would be evidently susceptible of infinite variation.

Aberration and Astronomy.

It is possible also that astronomy may some day furnish us data on this point; she it was in the main who raised the question in making us acquainted with the phenomenon of the aberration of light. If we make crudely the theory of aberration, we reach a very curious result The apparent positions of the stars differ from their real positions because of the earth's motion, and as this motion is variable, these apparent positions vary. The real position we can not ascertain, but we can observe the variations of the apparent position. The observations of the aberration show us, therefore, not the earth's motion, but the variations of this motion; they can not, therefore, give us information about the absolute motion of the earth.

At least this is true in first approximation, but the case would be no longer the same if we could appreciate the thousandths of a second. Then it would be seen that the amplitude of the oscillation depends not alone on the variation of the motion, a variation which is well known, since it is the motion of our globe on its elliptic orbit, but on the mean value of this motion, so that the constant of aberration would not be quite the same for all the stars, and the differences would tell us the absolute motion of the earth in space.

This, then, would be, under another form, the ruin of the principle of relativity. We are far, it is true, from appreciating the thousandth of a second, but, after all, say some, the earth's total absolute velocity is perhaps much greater than its relative velocity with respect to the sun. If, for example, it were 300 kilometers per second in place of 30, this would suffice to make the phenomenon observable.

I believe that in reasoning thus one admits a too simple theory of aberration. Michelson has shown us, I have told you, that physical procedures are powerless to put in evidence absolute motion; I am persuaded that the same will be true of the astronomic procedures, however far precision be carried.

However that may be, the data astronomy will furnish us in this regard will some day be precious to the physicist. Meanwhile, I believe that the theorists, recalling the experience of Michelson, may anticipate a negative result, and that they would accomplish a useful work in constructing a theory of aberration which would explain this in advance.

Electrons and Spectra.

This dynamics of electrons can be approached from many sides, but among the ways leading thither is one which has been somewhat neglected, and yet this is one of those which promise us the most surprises. It is movements of electrons which produce the lines of the emission spectra; this is proved by the Zeeman effect; in an incandescent body what vibrates is sensitive to the magnet, therefore electrified. This is a very important first point, but no one has gone farther. Why are the lines of the spectrum distributed in accordance with a regular law? These laws have been studied by the experimenters in their least details; they are very precise and comparatively simple. A first study of these distributions recalls the harmonics encountered in acoustics; but the difference is great. Not only are the numbers of vibrations not the successive multiples of a single number, but we do not even find anything analogous to the roots of those transcendental equations to which we are led by so many problems of mathematical physics: that of the vibrations of an elastic body of any form, that of the Hertzian oscillations in a generator of any form, the problem of Fourier for the cooling of a solid body.

The laws are simpler, but they are of wholly other nature, and to cite only one of these differences, for the harmonics of high order, the number of vibrations tends toward a finite limit, instead of increasing indefinitely.

That has not yet been accounted for, and I believe that there we have one of the most important secrets of nature. A Japanese physicist, M. Nagaoka, has recently proposed an explanation; according to him, atoms are composed of a large positive electron surrounded by a ring formed of a great number of very small negative electrons. Such is the planet Saturn with its rings. This is a very interesting attempt, but not yet wholly satisfactory; this attempt should be renewed. We will penetrate, so to speak, into the inmost recess of matter. And from the particular point of view which we to-day occupy, when we know why the vibrations of incandescent bodies differ thus from ordinary elastic vibrations, why the electrons do not behave like the matter which is familiar to us, we shall better comprehend the dynamics of electrons and it will be perhaps more easy for us to reconcile it with the principles.

Conventions Preceding Experiment.

Suppose, now, that all these efforts fail, and, after all, I do not believe they will, what must be done? Will it be necessary to seek to mend the broken principles by giving what we French call a *coup de pouce*? That evidently is always possible, and I retract nothing of what I have said above.

Have you not written, you might say if you wished to seek a quarrel with me — have you not written that the principles, though of experimental origin, are now unassailable by experiment because they have become conventions? And now you have just told us that the most recent conquests of experiment put these principles in danger.

Well, formerly I was right and to-day I am not wrong. Formerly I was right, and what is now happening is a new proof of it. Take, for example, the calorimetric experiment of Curie on radium. Is it possible to reconcile it with the principle of the conservation of energy? This has been attempted in many ways. But there is among them one I should like you to notice; this is not the explanation which tends to-day to prevail, but it is one of those which have been proposed. It has been conjectured that radium was only an intermediary, that it only stored radiations of unknown nature which flashed through space in every direction, traversing all bodies, save radium, without being altered by this passage and without exercising any action upon them. Radium alone took from them a little of their energy and afterward gave it out to us in various forms.

What an advantageous explanation, and how convenient! First, it is unverifiable and thus irrefutable. Then again it will serve to account for any derogation whatever to Mayer's principle; it answers in advance not only the objection of Curie, but all the objections that future experimenters might accumulate. This new and unknown energy would serve for everything.

This is just what I said, and therewith we are shown that our principle is unassailable by experiment.

But then, what have we gained by this stroke? The principle is intact, but thenceforth of what use is it? It enabled us to foresee that in such or such circumstance we could count on such a total quantity of energy; it limited us; but now that this indefinite provision of new energy is placed at our disposal, we are no longer limited by anything; and, as I have written in 'Science and Hypothesis,' if a principle ceases to be fecund, experiment without contradicting it directly will nevertheless have condemned it.

Future Mathematical Physics.

This, therefore, is not what would have to be done; it would be necessary to rebuild anew. If we were reduced to this necessity, we could moreover console ourselves. It would not be necessary thence to conclude that science can weave only a Penelope's web, that it can raise only ephemeral structures, which it is soon forced to demolish from top to bottom with its own hands.

As I have said, we have already passed through a like crisis. I have shown you that in the second mathematical physics, that of the principles, we find traces of the first, that of central forces; it will be just the same if we must know a third. Just so with the animal that exuviates, that breaks its too narrow carapace and makes itself a fresh one; under the new envelope one will recognize the essential traits of the organism which have persisted.

We can not foresee in what way we are about to expand; perhaps it is the kinetic theory of gases which is about to undergo development and serve as model to the others. Then the facts which first appeared to us as simple thereafter would be merely resultants of a very great number of elementary facts which only the laws of chance would make cooperate for a common end. Physical law would then assume an entirely new aspect; it would no longer be solely a differential equation, it would take the character of a statistical law.

Perhaps, too, we shall have to construct an entirely new mechanics that we only succeed in catching a glimpse of, where, inertia increasing with the velocity, the velocity of light would become an impassable limit. The ordinary mechanics, more simple, would remain a first approximation, since it would be true for velocities not too great, so that the old dynamics would still be found under the new. We should not have to regret having believed in the principles, and even, since velocities too great for the old formulas would always be only exceptional, the surest way in practise would be still to act as if we continued to believe in them. They are so useful, it would be necessary to keep a place for them. To determine to exclude them altogether would be to deprive oneself of a precious weapon. I hasten to say in conclusion that we are not yet there, and as yet nothing proves that the principles will not come forth from out the fray victorious and intact. ([1] These considerations on mathematical physics are borrowed from my St. Louis address.)

On the Dynamics of the Electron (1905)
by Henri Poincaré, translated from French by Wikisource

ELECTRICITY —— On the dynamics of the electron.

Note of M. H. POINCARÉ

It seems at first sight that the aberration of light and the associated optical phenomena will provide a means of determining the absolute motion of the Earth, or rather its motion, not in relation to other stars, but in relation to the ether. This is not the case: the experiments in which we take into account only the first order of aberration were initially unsuccessful and an explanation was easily found; but MICHELSON, who imagined an experiment by which terms depending on the square of the aberration could be measured, had no luck either. It seems that this inability to demonstrate absolute motion is a general law of nature.

An explanation was proposed by LORENTZ, who introduced the hypothesis of a contraction of all bodies in the direction of motion of the earth; this contraction would account for the Michelson-Morley experiment and all those that have been conducted to date, but leaves room for other experiments even more delicate and more easily conceived than executed, which might demonstrate the absolute motion of the Earth. But if the impossibility of such a finding is considered highly probable, one can predict that these experiments, if they can ever be conducted, will give a negative result. LORENTZ has sought to supplement and amend his hypothesis so as to bring it into accord with the postulate of the *complete* impossibility of determining absolute motion. This he managed to do in his article entitled *Electromagnetic phenomena in a system moving with any velocity smaller than that of Light* (*Proceedings* de l'Académie d'Amsterdam, May 27, 1904).

The importance of this question made me determined to return to it; and the results I obtained are in agreement on all important points with those of LORENTZ; I was only led to modify and complete them in a few points of detail.

The essential point, established by LORENTZ, is that the electromagnetic field equations are not altered by a certain transformation (which I shall call after the name of *Lorentz*), which has the following form:

(1) $$x' = kl(x + \epsilon t), \quad y' = ly, \quad z' = lz, \quad t' = kl(t + \epsilon x)$$

x, y, z are the coordinates and t the time before the transformation, x', y', z' and t' after the transformation. Moreover, ϵ is a constant which defines the transformation

$$k = \frac{1}{\sqrt{1 - \epsilon^2}},$$

and l is an arbitrary function of ϵ. One can see that in this transformation the x-axis plays a particular role, but one can obviously construct a transformation in which this role would be played by any straight line through the origin. The sum of all these transformations, together with the set of all rotations of space, must form a group; but for this to occur, we need $l = 1$; so one is forced to suppose $l = 1$ and this is a consequence that LORENTZ has obtained by another way.

Let ρ the electric density of the electron, ξ, η, ζ the velocity before the transformation; we obtain for the same quantities $\rho', \xi', \eta', \zeta'$ after processing

(2) $$\rho' = \frac{k}{l^3}\rho(1 + \epsilon\xi), \quad \rho'\xi' = \frac{k}{l^3}\rho(\xi + \epsilon), \quad \rho'\eta' = \frac{\rho\eta}{l^3}, \quad \rho'\zeta' = \frac{\rho\zeta}{l^3}$$

These formulas differ somewhat from those which had been found by LORENTZ.

Let now X, Y, Z and X', Y', Z' the three components of force before and after transformation, and the force is expressed in unit volume; I found

(3) $$X' = \frac{k}{l^3}(X + \epsilon\Sigma X\xi), \quad Y' = \frac{Y}{l^3}, \quad Z' = \frac{Z}{l^3}$$

These formulas are slightly different from those of LORENTZ; the additional term in $\Sigma X\xi$ reminds us on a result previously obtained by LIÉNARD.

If we now denote by X_1, Y_1, Z_1 and X'_1, Y'_1, Z'_1 the components of a force, not referred to unit volume, but to unit mass of the electron, we obtain

(4) $$X'_1 = \frac{k}{l^3}\frac{\rho}{\rho'}(X_1 + \epsilon\Sigma X_1\xi), \quad Y'_1 = \frac{\rho}{\rho'}\frac{Y_1}{l^3}, \quad Z'_1 = \frac{\rho}{\rho'}\frac{Z_1}{l^3}$$

LORENTZ was also led to assume that the electron in motion takes the form of an oblate spheroid; this is also the hypothesis made by LANGEVIN, however, while LORENTZ assumed that two axes of the ellipsoid remain constant, which is consistent with the hypothesis $l = 1$, LANGEVIN assumed that the volume remains constant. Both authors have shown that these two hypotheses are consistent with the experiments of KAUFMANN, as well as the original hypothesis of ABRAHAM (spherical electron). The hypothesis of LANGEVIN would have the advantage that it is self-sufficient, because it suffices to regard the electron as deformable and incompressible, and to explain that it takes an ellipsoidal shape when it moves. But I show, in agreement with LORENTZ, that it is incapable to accord with the impossibility of an experiment showing the absolute motion. As I have said, this is because $l = 1$ is the only case for which all the Lorentz transformations form a group.

But with the hypothesis of LORENTZ, the agreement between the formulas does not occur all alone; we obtain at the same time a possible explanation for the contraction of the electron, assuming that *the electron, deformable and compressible, is subjected to a constant external pressure whose work is proportional to volume changes.*

I show, by applying the principle of least action, that under those conditions the compensation is complete, assuming that inertia is an electromagnetic phenomenon exclusively, as generally admitted since KAUFMANN's experiments, and apart from the constant pressure that I just mentioned and which acts on the electron, all the forces are of electromagnetic origin. We have thus the explanation of the impossibility of demonstrating absolute motion and of the contraction of all the bodies in the direction of the terrestrial motion.

But that's not all: LORENTZ, in the work quoted, found it necessary to complete his hypothesis by assuming that all forces, whatever their origin, are affected by translation in the same way as electromagnetic forces and, consequently, the effect produced on their components by the Lorentz transformation is still defined by equations (4).

It was important to examine this hypothesis more closely and in particular to examine what changes it would require us to make on the law of gravitation. That is what I sought to determine; I was first led to suppose that the propagation of gravitation is not instantaneous, but happens with the speed of light. This seems at odds with results obtained byLAPLACE, who announced that this propagation is, if not instantaneous, at least much faster than that of light. But in reality, the question posed by LAPLACE differs considerably from that which occupies us here. The introduction of a finite velocity of propagation was the only change Laplace introduced to NEWTON's law. Here, on the contrary, this change is accompanied by several others; it is possible, and that is indeed what happens, that a partial compensation occurs between them.

When we therefore speak of the position or velocity of the attracting body, it will be the position or the velocity at the time when the *gravitational wave* leaves the body; when we talk about the position or velocity of the attracted body, it will be the position or the velocity at the moment when this body was reached and attracted by the gravitational wave emanating from the other body; it is clear that the first instant precedes the second.

So if x, y, z are the projections on the three axes of the vector joining the two positions, if the velocity of the attracted body is ξ, η, ζ, and that of the attracting body ξ_1, η_1, ζ_1, the three components of the attraction (which I can still call X'_1, Y'_1, Z'_1) are functions of $x, y, z, \xi, \eta, \zeta, \xi_1, \eta_1, \zeta_1$. I asked myself whether it was possible to determine these functions in a way that they are affected by the Lorentz transformation according to equations (4) and found the ordinary law of gravitation, whenever the velocities $\xi, \eta, \zeta, \xi_1, \eta_1, \zeta_1$ are small enough that one can neglect the squares in respect to the square of the speed of light.

The answer must be affirmative. It is found that the corrected attraction consists of two forces, one parallel to the vector x, y, z, the other to the velocity ξ, η, ζ.

The difference to the ordinary law of gravitation, as I have said, is of order ξ^2; if we only assume, as LAPLACE did, that the speed of propagation is that of light, this discrepancy is of order ξ, that is to say 10.000 times larger. It is therefore, at first sight, not absurd to assume that astronomical observations are not precise enough to detect a difference as small as the one which we imagine. But this is what only a thorough discussion will make possible to decide.

On the Dynamics of the Electron

By **H. Poincaré** (Paris).

Meeting of July 23, 1905.

Introduction.

It seems at first sight that the aberration of light and the related optical and electrical phenomena will provide us a means of determining the absolute motion of the Earth, or rather its motion, not in relation to the other stars, but in relation to the ether. FRESNEL had already tried it, but he recognized soon that the motion of the earth does not alter the laws of refraction and reflection. Similar experiments, like that of a telescope filled with water and all those which take into consideration only terms of first order in respect to aberration, give no other but negative results; soon an explanation was discovered; but MICHELSON, having imagined an experiment where the terms depending on the square of the aberration became sensitive, failed as well.

It seems that this impossibility of demonstrating an experimental evidence for absolute motion of the Earth is a general law of nature; we are naturally led to admit this law, which we will call the *Postulate of Relativity* and admit it without restriction. This postulate, which is up to now in accord with experiments, may be either confirmed or disproved later by more precise experiments, it is in any case interesting to see which consequences follow from it.

An explanation was proposed by LORENTZ and FITZGERALD, who introduced the hypothesis of a contraction undergone by all bodies into the direction of the motion of earth and proportional to the square of aberration; this contraction, which we will call *LORENTZ contraction*, would give an account of the **experiment of** MICHELSON and all those which were carried out up to now. The hypothesis would become insufficient, however, if one were to assume the postulate of relativity in all its generality.

LORENTZ sought to supplement and modify it in order to put it in perfect agreement with this postulate. He succeeded in doing so in his article entitled *Electromagnetic phenomena in a system moving with any velocity smaller than that of light* (*Proceedings* de l'Académie d'Amsterdam, May 27, 1904).

The importance of the question determined me to take it up again; the results which I obtained are in agreement with those of LORENTZ on all important points; I was only led to modify and supplement them in some points of detail; one will further see the differences which are of secondary importance.

The idea of LORENTZ can be summarized as follows: if we can bring the whole system to a common translation, without modification of any of the apparent phenomena, it is because the equations of the electromagnetic medium are not altered by certain transformations, which we will call *LORENTZ transformation*; two systems, one motionless, the other in translation, thus become exact images of one another.

LANGEVIN[1] had sought to modify the idea of LORENTZ; for both authors the moving electron takes the shape of a flattened ellipsoid, but for LORENTZ two of the axes of the ellipsoid remain constant, while for LANGEVIN on the contrary it is the volume of the ellipsoid which remains constant. Besides, both scientists showed hat these two hypothesis are in agreement with the experiments of KAUFMANN, as well as the original hypothesis of Abraham (undeformable spherical electron).

The advantage of the theory of LANGEVIN is that it uses only electromagnetic forces and binding forces; but it is incompatible with the postulate of relativity; this is what LORENTZhad shown, this is what I find again in another way by relying upon the principles of group theory.

It is thus necessary to return from here to the theory of LORENTZ; but if one wants to preserve it and avoid intolerable contradictions, it is necessary to suppose a special force which explains at the same time the contraction and the constancy of two of the axes. I sought to determine this force, I found that *it can be compared to a constant external pressure, acting on the deformable and compressible electron, and whose work is proportional to the variations of the volume of the electron.*

So if the inertia of matter is exclusively of electromagnetic origin, as it is generally admitted since the experiment of KAUFMANN, and except that constant pressure from which I come to speak, all forces are of electromagnetic origin, the postulate of relativity can be established in any rigour. It is what I show by a very simple calculation founded on the principle of least action.

But this is not all. LORENTZ, in the quoted work, considered it to be necessary to supplement his hypothesis so that the postulate remains when there are other forces as the electromagnetic forces. According to him, all the forces, whatever is their origin, are affected by the LORENTZ transformation (and consequently by a translation) in the same way as the electromagnetic forces.

It was important to examine this assumption more closely and in particular to seek which modifications it would oblige us to bring to the laws of gravitation.

It is found at first sight, that we are forced to suppose that the propagation of gravitation is not instantaneous, but happens with the speed of light. One could believe that this is a sufficient reason to reject the hypothesis, as LAPLACE has shown that this cannot be so. But actually, this propagation effect is mainly compensated by a different cause, so that there is no more contradiction between the proposed law and the astronomical observations.

Is it possible to find a law, which satisfies the condition imposed by LORENTZ, and which at the same time is reduced to the law of Newton when the speeds of the stars are rather small, so that one can neglect their squares (as well as the product of acceleration and distance) in respect to the square speed of light?

To this question, as it further will be seen, one must answer in the affirmative.

Is the law thus amended compatible with the astronomical observations?

At first sight it seems that it is the case, but this question can be decided only by a thorough discussion.

But even accepting that the discussion turns to the advantage of a new hypothesis, what should we conclude? If the propagation of attraction happens with the speed of light, it cannot be by a fortuitous coincidence, it must be due to a function of the ether; and then it will be necessary to seek to penetrate the nature of this function, and to relate it to the other functions of the fluid.

We cannot be satisfied with simply juxtaposed formulas which would agree only by a lucky stroke; it is necessary that these formulas are so to speak able to be penetrated mutually. Our mind will not be satisfied before it believes to see the reason of this agreement, at the point where it has the illusion that it could have predicted it.

But the question can still be seen form another point of view, which could be better understood by analogy. Let us suppose an astronomer before COPERNICUS who reflects on the system of PTOLEMY; he will notice that for all planets one of the two circles, epicycle or deferent, is traversed in the same time. This cannot be by chance, there is thus between all planets a mysterious binding.

But COPERNICUS, by simply changing the axes of coordinates regarded as fixed, destroyed this appearance;

each planet does not describe any more than only one circle and the durations of the revolutions become independent (until KEPLER restores between them the binding which was believed to be destroyed).

Here it is possible that there is something analogue; if we admit the postulate of relativity, we would find in the law of gravitation and the electromagnetic laws a common number which would be the speed of light; and we would still find it in all the other forces of any origin, which could be explained only in two manners:

Either there would be nothing in the world which is not of electromagnetic origin.

Or this part which would be, so to speak, common to all the physical phenomena, would be only apparent, something which would be due to our methods of measurement. How do we perform our measurements? By transportation, one on the other, of objects regarded as invariable solids, one will answer immediately; but this is not true any more in the current theory, if the LORENTZ contraction is admitted. In this theory, two equal lengths are, by definition, two lengths for which light takes the same time to traverse.

Perhaps it would be enough to give up this definition, so that the theory of LORENTZ is as completely rejected as it was the system of PTOLEMY by the intervention of COPERNICUS. If that happens one day, it will not prove that the effort made by LORENTZ was useless; because PTOLEMY, no matter what we think about him, was not useless for COPERNICUS.

Also I did not hesitate to publish these few partial results, although in this moment even the whole theory seems to be endangered by the discovery of magnetocathodic rays.

§ 1. — LORENTZ transformation

LORENTZ had adopted a particular system of units, so as to eliminate the factors 4π in the formulas. I'll do the same, plus I choose the units of length and time so that the speed of light is equal to 1. Under these conditions the fundamental formulas become (by calling f, g, h the electric displacement, α, β, γ the magnetic force, F, G and H the vector potential, φ the scalar potential, ρ the electric density, ξ, η, ζ the electron velocity, u, v, w the current):

(1)
$$
\begin{cases}
u = \frac{df}{dt} + \rho\xi = \frac{d\gamma}{dy} - \frac{d\beta}{dz}, \quad \alpha = \frac{dH}{dy} - \frac{dG}{dz}, \quad f = -\frac{dF}{dt} - \frac{d\psi}{dx}, \\[2mm]
\frac{d\alpha}{dt} = \frac{dg}{dz} - \frac{dh}{dy}, \quad \frac{d\rho}{dt} + \sum \frac{d\rho\xi}{dx} = 0, \quad \sum \frac{df}{dx} = \rho, \quad \frac{d\psi}{dt} + \sum \frac{dF}{dx} = 0, \\[2mm]
\Box = \Delta - \frac{d^2}{dt^2} = \sum \frac{d^2}{dx^2} - \frac{d^2}{dt^2}, \quad \Box\psi = -\rho, \quad \Box F = -\rho\xi.
\end{cases}
$$

A material element of volume $dx\,dy\,dz$ suffers a mechanical force whose components $X\,dx\,dy\,dz$, $Y\,dz\,dx\,dy$, $Z\,dx\,dy\,dz$ are deduced from the formula:

(2)
$$ X = \rho f + \rho(\eta\gamma - \zeta\beta) $$

These equations are capable of a remarkable transformation discovered by LORENTZ and which owes its interest from the fact, that it explains why no experience is suited to show us the absolute motion of the universe. Let:

(3)
$$ x' = kl(x + \epsilon t), \quad t' = kl(t + \epsilon x), \quad y' = ly, \quad z' = lz $$

l and ϵ are two arbitrary constants, and

$$ k = \frac{1}{\sqrt{1 - \epsilon^2}}. $$

If we now set:

$$ \Box' = \sum \frac{d^2}{dx'^2} - \frac{d^2}{dt'^2}, $$

it follows:

$$ \Box' = \Box l^{-2}. $$

Consider a sphere entrained with the electron in a uniform translational motion, and

$$ (x - \xi t)^2 + (y - \eta t)^2 + (z - \zeta t)^2 = r^2, $$

is the equation of that moving sphere whose volume is $\frac{4}{3}\pi r^2$.

The transformation will change it into an ellipsoid, and it is easy to find the equation. It is easily deduced because of equations (3):

$$x = \frac{k}{l}(x' - \epsilon t'), \; t = \frac{k}{l}(t' - \epsilon x'), \; y = \frac{y'}{l}, \; z = \frac{z'}{l}.$$

The equation of the ellipsoid becomes:

$$k^2(x' - \epsilon t' - \xi t' + \epsilon \xi x')^2 + (y' - \eta k t' + \eta k \epsilon x')^2 + (z' - \zeta k t' + \zeta k \epsilon x')^2 = l^2 r^2.$$

This ellipsoid moves in uniform motion; for $t' = 0$, it reduces to

$$k^2 x'^2(1 + \xi \epsilon)^2 + (y' + \eta k \epsilon x')^2 + (z' + \zeta k \epsilon x')^2 = l^2 r^2,$$

and has the volume:

$$\frac{4}{3}\pi r^3 \frac{l^3}{k(1 + \xi \epsilon)}.$$

If we want that the charge of an electron is not altered by the transformation, and when we call ρ' the new electrical density, it follows:

(4)
$$\rho' = \frac{k}{l^3}(\rho + \epsilon \rho \xi).$$

Those are the new velocities ξ', η', ζ '; we must have:

$$\xi' = \frac{dx'}{dt'} = \frac{d(x + \epsilon t)}{d(t + \epsilon x)} = \frac{\xi + \epsilon}{1 + \epsilon \xi},$$

$$\eta' = \frac{dy'}{dt'} = \frac{dy}{kd(t + \epsilon x)} = \frac{\eta}{k(1 + \epsilon \xi)}, \quad \zeta' = \frac{\zeta}{k(1 + \epsilon \xi)},$$

where:

$$\rho'\xi' = \frac{k}{l^3}(\rho\xi + \epsilon\rho), \quad \rho'\eta' = \frac{1}{l^3}\rho\eta, \quad \rho'\zeta' = \frac{1}{l^3}\rho\zeta$$

Here I should mention for the first time a discrepancy with LORENTZ.

LORENTZ poses (with different notations) (loco citato, page 813, formulas 7 and 8):

$$\rho' = \frac{1}{kl^3}\rho, \quad \xi' = k^2(\xi + \epsilon), \quad \eta' = k\eta, \quad \zeta' = k\zeta.$$

We thus find the formulas:

$$\rho'\xi' = \frac{k}{l^3}(\rho\xi + \epsilon\rho), \quad \rho'\eta' = \frac{1}{l^3}\rho\eta, \quad \rho'\zeta' = \frac{1}{l^3}\rho\zeta;$$

but the value of ρ' differs.

It is important to note that formulas (4) and (4bis) satisfy the continuity condition

$$\frac{d\rho'}{dt'} + \sum \frac{d\rho'\xi'}{dx'} = 0.$$

Indeed, let λ be an undetermined quantity and D the functional determinant

(5)
$$t + \lambda\rho, \quad x + \lambda\rho\xi, \quad x + \lambda\rho\eta, \quad z + \lambda\rho\zeta$$

with respect to t, x, y, z. We will have:

$$D = D_0 + D_1\lambda + D_2\lambda^2 + D_3\lambda^3 + D_4\lambda^4$$

$$\text{with} \quad D_0 = 1, \ D_1 = \frac{d\rho}{dt} + \sum \frac{d\rho\xi}{dx} = 0.$$

Let $\lambda' = l^2\lambda$, we see that the four functions

5bis
$$t' + \lambda'\rho', \quad x' + \lambda'\rho'\xi', \quad y' + \lambda'\rho'\eta', \quad z' + \lambda'\rho'\zeta'$$

are related to the functions (5) by the same linear relations as the old variables to the new variables. Then, if we denote by D' the functional determinant of the functions (5bis) in relation to the new variables, we have:

$$D' = D, \quad D' = D'_0 + D'_1\lambda' + \ldots + D'_4\lambda'^4,$$

where:

$$D'_0 = D_0 = 1, \quad D'_1 = l^{-2}D_1 = 0 = \frac{d\rho'}{dt'} + \sum \frac{d\rho'\xi'}{dx'}. \quad \text{C. Q. F. D}$$

With the hypothesis of LORENTZ, this condition is not satisfied, since ρ' has not the same value.

We will define the new potentials, vector and scalar, in order to satisfy the conditions

(6)
$$\Box'\psi' = -\rho', \quad \Box'F' = -\rho'\xi'.$$

Then we obtain from this:

(7)
$$\psi' = \frac{k}{l}(\psi + \epsilon F), \ F' = \frac{k}{l}(F + \epsilon\psi), \ G' = \frac{1}{l}G, \ H' = \frac{1}{l}H.$$

These formulas differ significantly from those of LORENTZ, but the difference is ultimately due to the definitions.

We will choose the new electric and magnetic fields so as to satisfy the equations:

(8)
$$f' = -\frac{dF'}{dt'} - \frac{d\psi'}{dx'}, \quad \alpha' = \frac{dH'}{dy'} - \frac{dG'}{dz'}.$$

It is easy to see that:

$$\frac{d}{dt'} = \frac{k}{l}\left(\frac{d}{dt} - \epsilon\frac{d}{dx}\right), \quad \frac{d}{dx'} = \frac{k}{l}\left(\frac{d}{dx} - \epsilon\frac{d}{dt}\right), \quad \frac{d}{dy'} = \frac{1}{l}\frac{d}{dy}, \quad \frac{d}{dz'} = \frac{1}{l}\frac{d}{dz}$$

and we conclude:

(9)
$$\begin{cases} f' = \frac{1}{l^2}f, g' = \frac{k}{l^2}(g + \epsilon\gamma), \quad h' = \frac{k}{l^2}(h - \epsilon\beta), \\ \alpha' = \frac{1}{l^2}\alpha, \beta' = \frac{k}{l^2}(\beta - \epsilon h), \quad \gamma' = \frac{k}{l^2}(\gamma + \epsilon g). \end{cases}$$

These formulas are identical to those of LORENTZ.

Our transformation does not alter the equations (I). Indeed, the continuity condition, and the equations (6) and (8), already provided us with some of the equations (I) (except the accentuation of letters).

Equations (6) close to the continuity condition give:

(10)
$$\frac{d\psi'}{dt'} + \sum \frac{dF'}{dx'} = 0.$$

It remains to establish that:

$$\frac{df'}{dt'} + \rho'\xi' = \frac{d\gamma'}{dy'} - \frac{d\beta'}{dz'}, \quad \frac{dz'}{dt'} = \frac{dg'}{dz'} - \frac{dh'}{dy'}, \quad \sum \frac{df'}{dx'} = \rho'$$

and it is easy to see that these are necessary consequences of equations (6), (8) and (10).

We must now compare the force before and after transformation.

Let X, Y, Z be the force before, and X', Y', Z' the force after transformation, both related to unit volume. In order for X' to satisfy the same equations as before the transformation, we must have:

$$X' = \rho' f' + \rho'(\eta'\gamma' - \zeta'\beta'),$$
$$Y' = \rho' g' + \rho'(\zeta'\alpha' - \xi'\gamma'),$$
$$X' = \rho' h' + \rho'(\xi'\beta' - \eta'\alpha'),$$

or, replacing all quantities by their values (4), (4bis) and (9) and taking into account equations (2):

(11)
$$\begin{cases} X' = \dfrac{k}{l^5}\left(X + \epsilon \sum X\xi\right), \\[2mm] Y' = \dfrac{1}{l^5}Y, \\[2mm] Z' = \dfrac{1}{l^5}Z. \end{cases}$$

If we represent the components of the force X_1, Y_1, Z_1, not per unit volume, but per unit of electric charge of the electron, and X'_1, Y'_1, Z'_1 are the same quantities after the transformation, we would have:

$$X_1 = f + \eta\gamma - \zeta\beta, \quad X'_1 = f' + \eta'\gamma' - \zeta'\beta', \quad X = \rho X_1, \quad X' = \rho' X'_1$$

and we would have the equations

(11bis)
$$\begin{cases} X'_1 = \dfrac{k}{l^5}\dfrac{\rho}{\rho'}\left(X_1 + \epsilon \sum X_1\xi\right), \\[2mm] Y'_1 = \dfrac{1}{l^5}\dfrac{\rho}{\rho'}Y_1, \\[2mm] Z'_1 = \dfrac{1}{l^5}\dfrac{\rho}{\rho'}Z_1. \end{cases}$$

LORENTZ found [with different notation, page 813, formula (10)]:

$$
(11^{\text{ter}}) \quad
\begin{cases}
X_1 = l^2 X_1' - l^2 \epsilon (\eta' g' + \zeta' h'), \\[2mm]
Y_1 = \dfrac{l^2}{k} Y_1' + \dfrac{l^2 \epsilon}{k} \xi' g', \\[2mm]
Z_1 = \dfrac{l^2}{k} Z_1' + \dfrac{l^2 \epsilon}{k} \xi' h'.
\end{cases}
$$

Before going further, it is important to investigate the cause of this significant discrepancy. It is obvious that the formulas for ξ', η', ζ' are not the same, while the formulas for the electric and magnetic fields are the same.

If the inertia of electrons is exclusively of electromagnetic origin, if in addition they are subject only to forces of electromagnetic origin, the equilibrium condition requires that we have inside the electrons:

$$
X = Y = Z = 0.
$$

But in virtue of equations (11) those relations are equivalent to

$$
X' = Y' = Z' = 0.
$$

The equilibrium conditions of the electrons are not altered by the transformation.

Unfortunately, a hypothesis as simple as that is unacceptable. If, indeed, we assume $\xi = \eta = \zeta = 0$, the conditions $X = Y = Z = 0$ entrain $f = g = h = 0$, and consequently $\sum \dfrac{df}{dx} = 0$, *i.e.* $\rho = 0$. We arrive at similar results in the most general case. We must therefore admit that there are, in addition to electromagnetic forces, either other forces or bindings. It is necessary to search for conditions which must satisfy these forces or bindings, so that the equilibrium of the electron is not disturbed by the transformation. This will be the subject of a later paragraph.

§ 2. — Principle of least action

We know how Lorentz deduced his equations from the principle of least action. I will return to this question, even though I have nothing substantial to add to the analysis of Lorentz, because I prefer to present it in a slightly different form which will be useful for my purpose. I will pose:

(1)
$$J = \int dt \; d\tau \left[\frac{\sum f^2}{2} + \frac{\sum \alpha^2}{2} - \sum Fu \right],$$

assuming that f, α, F, u, etc., are subject to the following conditions and the ones deduced by symmetry:

(2)
$$\sum \frac{df}{dx} = \rho, \quad \alpha = \frac{dH}{dy} - \frac{dG}{dz}, \quad u = \frac{df}{dt} + \rho\xi.$$

Regarding the integral J, it must be extended:

1° in relation to the volume element $d\tau = dx \; dy \; dz$ over the whole space;

2° in relation to time t, over the interval between the limits $t = t_0$, $t = t_1$.

According to the principle of least action, the integral J must be a minimum, if one sets the various quantities which appear in:

1° the conditions (2);

2° the condition that the state of the system is determined by both limiting times $t = t_0$, $t = t_1$.

This last condition allows us to transform our integral by partial integration with respect to time. If we have indeed an integral of the form

$$\int dt \; d\tau A \frac{dB\delta C}{dt},$$

where C is a quantity that defines the system state and its variation δC, it will be equal to (by partial integration with respect to time):

$$\int d\tau \; | \; AB\delta C \; | \; \begin{matrix} t = t_1 \\ t = t_0 \end{matrix} - \int dt \; d\tau \frac{dA}{dt} dB\delta C.$$

Since the system state is determined by both limiting times, it is $\delta C = 0$ for $t = t_0$, $t = t_1$, so the first integral which is related to these two periods is zero, and the 2nd one remains.

We can also integrate by parts with respect to x, y or z, we have indeed

$$\int A \frac{dB}{dx} dx \; dy \; dz \; dt = \int AB \; dy \; dz \; dt - \int B \frac{dA}{dx} dx \; dy \; dz \; dt.$$

Our integrations are extended to infinity, it must be $x = \pm\infty$ in the first integral on the right-hand side; so, since we always assume that all our functions vanish at infinity, this integral will be zero and it follows

$$\int A \frac{dB}{dx} d\tau \; dt = -\int B \frac{dA}{dx} d\tau \; dt.$$

If the system is supposed to be subject to bindings, the binding conditions should be connected to the conditions imposed on the various quantities appearing in the integral J.

Let us first give to F, G, H the increasements δF, δG, δH; where:

$$\delta\alpha = \frac{d\delta H}{dy} - \frac{d\delta G}{dz}.$$

We should have

$$\delta J = \int dt \; d\tau \left[\sum \alpha \left(\frac{d\delta H}{dy} - \frac{d\delta G}{dz} \right) - \sum u\delta F \right] = 0,$$

or, integrating by parts,

$$\delta J = \int dt \; d\tau \left[\sum \left(\delta G \frac{d\alpha}{dz} - \delta H \frac{d\alpha}{dy} \right) - \sum u\delta F \right] =$$

$$= -\int dt \; d\tau \sum \delta F \left(u - \frac{d\gamma}{dy} + \frac{d\beta}{dz} \right) = 0$$

whence, by setting the arbitrary coefficient δF equal to zero,

(3) $$u = \frac{d\gamma}{dy} - \frac{d\beta}{dz}.$$

This relationship gives us (by partial integration):

$$\int \sum Fu d\tau = \int \sum F \left(\frac{d\gamma}{dy} - \frac{d\beta}{dz} \right) d\tau = \int \sum \left(\beta \frac{dF}{dz} - \gamma \frac{dF}{dy} \right) d\tau =$$

$$= \int \sum \alpha \left(\frac{dH}{dy} - \frac{dG}{dz} \right) d\tau,$$

or

$$\int \sum Fu d\tau = \int \sum \alpha^2 d\tau,$$

hence finally:

(4) $$J = \int dt \; d\tau \left(\frac{\sum f^2}{2} - \frac{\sum \alpha^2}{2} \right).$$

Now, thanks to equation (3), δJ is independent from δF and thus δα; let us vary now the other variables

It follows, by returning to expression (1) of J,

$$\delta J = \int dt \; d\tau \left(\sum f \delta f - \sum F \delta u \right).$$

But f, g, h are first subject to conditions (2), so that

(5)
$$\sum \frac{d\delta f}{dx} = \delta \rho$$

and for convenience we write:

(6)
$$\delta J = \int dt \; d\tau \left[\sum f df - \sum F \delta u - \psi \left(\sum \frac{d\delta f}{dx} - \delta \rho \right) \right].$$

The principles of variation calculus tells us that we must do the calculation as if ψ is an arbitrary function, as if δJ is represented by (6), and as if the changes were no longer subject to the condition (5).

We have in addition:

$$\delta u = \frac{d\delta f}{dt} + \delta \rho \xi,$$

whence, after partial integration,

(7)
$$\delta J = \int dt \; d\tau \sum \delta f \left(f + \frac{dF}{dt} + \frac{d\psi}{dx} \right) + \int dt \; d\tau \left(\psi \delta \rho - \sum F \delta \rho \xi \right).$$

If we assume at first that the electrons do not undergo a variation, $\delta \rho = \delta \rho \xi = 0$ and the second integral is zero. Because δJ must vanish, we should have:

(8)
$$f + \frac{dF}{dt} + \frac{d\psi}{dx} = 0$$

It remains in the general case:

(9)
$$\delta J = \int dt \; d\tau \left(\psi \delta \rho - \sum F \delta \rho \xi \right).$$

It remains to determine the forces acting on the electrons. To do this we must suppose that a supplementary force -X$d\tau$, -Y$d\tau$, -Z$d\tau$ applies to each element of an electron, and write that this force is in equilibrium with the forces of electromagnetic origin. Let U, V, W be components of the displacement of the element $d\tau$ of the electron, where the displacement is counted from an arbitrary initial position. Let δU, δV, δW be the variations of this displacement; the virtual work corresponding to the supplementary force is:

$$- \int \sum X \delta U \; d\tau,$$

so that the equilibrium condition about which we have spoken can be written:

(10)
$$\delta J = - \int \sum X \delta U \; d\tau \; dt.$$

It's about the transformation of δJ. To begin the search for the continuity equation, we express how the charge of an electron is preserved by the variation.

Let x_0, y_0, z_0 be the initial position of an electron. Its current position is:

$$x = x_0 + U, \quad y = y_0 + V, \quad z = z_0 + W.$$

We also introduce an auxiliary variable ε, which produces changes in our various functions, so that for any function A we have:

$$\delta A = \delta\epsilon \frac{dA}{d\epsilon}$$

It is indeed convenient to switch from the notation of variation calculus to that of ordinary calculus, or vice versa.

Our functions should be regarded: 1° as dependent on five variables x, y, z, t, ε, so that we can remain at the same place when ε and t vary alone: we then indicate their derivatives by the ordinary d; 2° as dependent on five variables x_0, y_0, z_0, t, ε so that we may always follow a single electron when t and ε vary alone, then we denote their derivatives by ∂. We will have then:

(11)
$$\xi = \frac{\partial U}{\partial t} = \frac{dU}{dt} + \xi \frac{dU}{dx} + \eta \frac{dU}{dy} + \zeta \frac{dU}{dz} = \frac{\partial x}{\partial t}.$$

Denote now by Δ the functional determinant of x, y, z with respect to x_0, y_0, z_0:

$$\Delta = \frac{\partial(x,\ y,\ z)}{\partial(x_0,\ y_0,\ z_0)}.$$

If ε, x_0, y_0, z_0 remain constant, we give to t an increasement ∂t; to x, y, z the increasements ∂x_0, ∂y_0, ∂z_0 will result; and to Δ the increasement $\partial \Delta$, and there will be:

$$\partial x = \xi \partial t, \quad \partial y = \eta \partial t, \quad \partial z = \zeta \partial t,$$
$$\Delta + \partial\Delta = \frac{\partial(x + \partial x,\ y + \partial y,\ z + \partial z)}{\partial(x_0,\ y_0,\ z_0)}$$

hence

$$1 + \frac{\partial\Delta}{\Delta} = \frac{\partial(x + \partial x,\ y + \partial y,\ z + \partial z)}{\partial(x,\ y,\ z)} = \frac{\partial(x + \xi\partial t,\ y + \eta\partial t,\ z + \zeta\partial t)}{\partial(x,\ y,\ z)}$$

We deduce:

(12)
$$\frac{1}{\Delta}\frac{\partial\Delta}{\partial t} = \frac{d\xi}{dx} + \frac{d\eta}{dy} + \frac{d\zeta}{dz}$$

The mass of each electron is invariable, we have:

(13)
$$\frac{\partial\rho\Delta}{\partial t} = 0$$

where:

$$\frac{\partial \rho}{\partial t} + \sum \rho \frac{d\xi}{dx} = 0, \quad \frac{\partial \rho}{\partial t} = \frac{d\rho}{dt} + \sum \xi \frac{d\rho}{dx}, \quad \frac{d\rho}{dt} + \sum \frac{d\rho\xi}{dx} = 0.$$

These are the different forms of the continuity equation with respect of variable t. We find similar forms with respect to the variable ε. Either:

$$\delta U = \frac{\partial U}{\partial \epsilon} \delta\epsilon, \quad \delta V = \frac{\partial V}{\partial \epsilon} \delta\epsilon, \quad \delta W = \frac{\partial W}{\partial \epsilon} \delta\epsilon;$$

it follows:

(11bis)
$$\delta U = \frac{dU}{d\epsilon}\delta\epsilon + \delta U \frac{dU}{dx} + \delta V = \frac{dU}{dy} + \delta W \frac{dU}{dz},$$

(12bis)
$$\frac{1}{\Delta}\frac{\partial \Delta}{\partial \epsilon} = \sum \frac{dU}{d\epsilon}, \quad \frac{\partial \rho\Delta}{\partial \epsilon} = 0,$$

(13bis)
$$\delta\epsilon \frac{\partial \rho}{\partial \epsilon} + \sum \rho \frac{d\rho U}{dx} = 0, \quad \frac{\partial \rho}{\partial \epsilon} = \frac{d\rho}{d\epsilon} + \sum \frac{\delta U}{\delta\epsilon}\frac{d\rho}{dx}, \quad \delta\rho + \frac{d\rho\delta U}{dx} = 0.$$

Note the difference between the definition of $\delta U = \frac{dU}{d\epsilon}\delta\epsilon$ and that of $\delta\rho = \frac{d\rho}{d\epsilon}\delta\epsilon$, we note that it is this definition of δU that suits to formula (10).

This equation will allow us to transform the first term of (9); we find in fact:

$$\int dt \; d\tau \; \psi\delta\rho = -\int dt \; d\tau \; \psi \sum \frac{d\rho\delta U}{dx},$$

or, by partial integration,

(14bis)
$$\int dt \; d\tau \; \psi\delta\rho = \int dt \; d\tau \sum \rho \frac{d\psi}{dx}\delta U.$$

Let us propose now to determine

$$\delta(\rho\xi) = \frac{d(\rho\xi)}{d\epsilon}\delta\epsilon$$

Note that ρΔ does not depend on x_0, y_0, z_0; indeed, if we consider an electron whose initial position is a rectangular parallelepiped whose edges are dx_0, dy_0, dz_0, the charge of this element is

$$\rho\Delta dx_0 \; dy_0 \; dz_0$$

and this charge should remain constant, then:

(15)
$$\frac{\partial \rho\Delta}{\partial t} = \frac{\partial \rho\Delta}{\partial \epsilon} = 0.$$

We deduce:

(16)
$$\frac{\partial^2 \rho\Delta U}{\partial t \; \partial \epsilon} = \frac{\partial}{\partial \epsilon}\left(\rho\Delta \frac{\partial U}{\partial t}\right) = \frac{\partial}{\partial t}\left(\rho\Delta \frac{\partial U}{\partial \epsilon}\right).$$

Now we know that for any function A, we have by the continuity equation,

$$\frac{1}{\Delta}\frac{\partial A\Delta}{\partial t} = \frac{dA}{dt} + \sum \frac{dA\xi}{dx}$$

and also

$$\frac{1}{\Delta}\frac{\partial A\Delta}{\partial \epsilon} = \frac{dA}{d\epsilon} + \sum \frac{dA\frac{\partial U}{\partial \epsilon}}{dx}$$

We thus have:

(17) $\quad \dfrac{1}{\Delta}\dfrac{\partial}{\partial \epsilon}\left(\rho\Delta\dfrac{\partial U}{\partial t}\right) = \dfrac{d\rho\frac{\partial U}{\partial t}}{d\epsilon} + \dfrac{d\left(\rho\frac{\partial U}{\partial t}\frac{\partial U}{\partial \epsilon}\right)}{dx} + \dfrac{d\left(\rho\frac{\partial U}{\partial t}\frac{\partial V}{\partial \epsilon}\right)}{dy} + \dfrac{d\left(\rho\frac{\partial U}{\partial t}\frac{\partial W}{\partial \epsilon}\right)}{dz},$

(17bis) $\quad \dfrac{1}{\Delta}\dfrac{\partial}{\partial t}\left(\rho\Delta\dfrac{\partial U}{\partial \epsilon}\right) = \dfrac{d\rho\frac{\partial U}{\partial \epsilon}}{dt} + \dfrac{d\left(\rho\frac{\partial U}{\partial t}\frac{\partial U}{\partial \epsilon}\right)}{dx} + \dfrac{d\left(\rho\frac{\partial V}{\partial t}\frac{\partial U}{\partial \epsilon}\right)}{dy} + \dfrac{d\left(\rho\frac{\partial W}{\partial t}\frac{\partial U}{\partial \epsilon}\right)}{dz}.$

The right-hand sides of (17) and (17bis) must be equal, and if one remembers that

$$\frac{\partial U}{\partial t} = \xi, \quad \frac{\partial U}{\partial \epsilon}\delta\epsilon = \delta U, \quad \frac{d\rho\xi}{d\epsilon}\delta\epsilon = \delta\rho\xi$$

we get:

(18) $\delta\rho\xi + \dfrac{d(\rho\xi\delta U)}{dx} + \dfrac{d(\rho\xi\delta V)}{dy} + \dfrac{d(\rho\xi\delta W)}{dz} = \dfrac{d(\rho\delta U)}{dt} + \dfrac{d(\rho\xi\delta U)}{dx} + \dfrac{d(\rho\eta\delta U)}{dy} + \dfrac{d(}{}$

Transforming now the second term of (9); we get:

$$\int dt\, d\tau \sum F\delta\rho\xi$$

$$= \int dt\, d\tau \left[\sum F\frac{d(\rho\delta U)}{dt} + \sum F\frac{d(\rho\eta\delta U)}{dy} + \sum F\frac{d(\rho\zeta\delta U)}{dz} - \sum F\frac{d(\rho\xi\delta V)}{dy}\right.$$

The right-hand side becomes by partial integration:

$$\int dt\, d\tau \left[-\sum \rho\delta U\frac{dF}{dt} - \sum \rho\eta\delta U\frac{dF}{dy} - \sum \rho\zeta\delta U\frac{dF}{dz} + \sum \rho\xi\delta V\frac{dF}{dy} + \sum \rho\xi\delta W\frac{dF}{dz}\right]$$

Now note, that:

$$\sum \rho\xi\delta V\frac{dF}{dy} = \sum \rho\zeta\delta U\frac{dH}{dx}, \quad \sum \rho\xi\delta W\frac{dF}{dz} = \sum \rho\eta\delta U\frac{dG}{dx}$$

If, indeed, we develop Σ on the two sides of these relations, they become identities; and remember that

$$\frac{dH}{dx} - \frac{dF}{dz} = -\beta, \quad \frac{dG}{dx} - \frac{dF}{dy} = \gamma$$

the right-hand side in question will become:

$$\int dt \ d\tau \left[-\sum \rho \delta U \frac{dF}{dt} - \sum \rho \gamma \eta \delta U - \sum \rho \beta \zeta \delta U \right],$$

so that finally:

$$\delta J = \int dt \ d\tau \sum \rho \delta U \left(\frac{d\psi}{dx} + \frac{dF}{dt} + \beta \zeta - \gamma \eta \right) = \int dt \ d\tau \sum \rho \delta U \left(-f + \beta \zeta - \gamma \eta \right)$$

Equating the coefficient of δU on both sides of (10) we get:

$$X = f - \beta \zeta + \gamma \eta$$

This is equation (2) of the preceding §.

§ 3. — The LORENTZ transformation and the principle of least action

Let us see if the principle of least action gives us the reason for the success of the LORENTZ transformation. We must look at the transformation of the integral:

$$J = \int dt \ d\tau \left(\frac{\sum f^2}{2} - \frac{\sum \alpha^2}{2} \right)$$

(formula 4 of § 2).

We first find

$$dt' \ d\tau' = l^4 dt \ d\tau$$

because x', y', z', t' are related to x, y, z, t by linear relations whose determinant is equal to l^4; then we have:

$$(1) \quad \begin{cases} l^4 \sum f'^2 = f^2 + k^2(g^2 + h^2) + k^2\epsilon^2(\beta^2 + \gamma^2) + 2k^2\epsilon(g\gamma - h\beta) \\ l^4 \sum \alpha'^2 = \alpha^2 + k^2(\beta^2 + \gamma^2) + k^2\epsilon^2(g^2 + h^2) + 2k^2\epsilon(g\gamma - h\beta) \end{cases}$$

(formula 9 of § 1), hence:

$$l^4 \left(\sum f'^2 - \sum \alpha'^2 \right) = \sum f^2 - \sum \alpha^2$$

so that if we set

$$J' = \int dt' \ d\tau' \left(\frac{\sum f'^2}{2} - \frac{\sum \alpha'^2}{2} \right),$$

we get:

$$J' = J.$$

However, to justify this equality, the integration limits have to be the same; so far we have assumed that t varies from t_0 to t_1, and x, y, z from ∞ to $+ \infty$. On this account the integration limits would be affected by the LORENTZ transformation, but nothing prevents us from assuming $t_0 =- \infty$, $t_1 = + \infty$; with those conditions the limits are the same for J and J'.

We then compare the following two equations analogues to equation (10) of § 2:

$$(2) \quad \begin{cases} \delta J = - \int \sum X \delta U \ d\tau \ dt \\ \delta J' = - \int \sum X' \delta U' \ d\tau' \ dt'. \end{cases}$$

For this, we must first compare δU' with δU.

Consider an electron whose initial coordinates are x_0, y_0, z_0; its coordinates at the instant t are

$$x = x_0 + U, \quad y = y_0 + V, \quad z = z_0 + W.$$

If one considers the electron after the corresponding LORENTZ transformation, it will have as coordinates

$$x' = kl\,(x + \epsilon t), \quad y' = ly, \quad z' = lz$$

where

$$x' = x_0 + U', \quad y' = y_0 + V', \quad z' = z_0 + W'$$

but it will only attain these coordinates at the instant

$$t' = kl\,(t + \epsilon x).$$

If we subject our variables to the variations δU, δV, δW, and when we give at the same time t an increasement δt, the coordinates x, y, z will experience a total increasement

$$\delta x = \delta U + \xi \delta t, \quad \delta y = \delta V + \eta \delta t, \quad \delta z = \delta W + \zeta \delta t.$$

We will also have:

$$\delta x' = \delta U' + \xi' \delta t', \quad \delta y' = \delta V' + \eta' \delta t', \quad \delta z' = \delta W' + \zeta' \delta t',$$

and in virtue of the LORENTZ transformation:

$$\delta x' = kl\,(\delta x + \epsilon \delta t), \quad \delta y' = l\delta y, \quad \delta z' = l\delta z, \quad \delta t' = kl\,(\delta t + \epsilon \delta x).$$

hence, assuming $\delta t = 0$, the relations:

$$\begin{cases} \delta x' = \delta U' + \xi' \delta t' & = kl\delta U, \\ \delta y' = \delta V' + \eta' \delta t' & = l\delta V, \\ \delta t' = kl\epsilon\delta U. \end{cases}$$

Note that

$$\xi' = \frac{\xi + \epsilon}{1 + \xi\epsilon}, \quad \eta' = \frac{\eta}{k\,(1 + \xi\epsilon)}$$

It follows, by replacing $\delta t'$ by its value

$$kl\,(1 + \xi\epsilon)\,\delta U = \delta U'\,(1 + \xi\epsilon) + (\xi + \epsilon)\,kl\epsilon\delta U,$$
$$l\,(1 + \xi\epsilon)\,\delta V = \delta V'\,(1 + \xi\epsilon) + \eta l\epsilon\delta U.$$

If we recall the definition of k, we draw from this:

$$\delta U = \frac{k}{l}\delta U' + \frac{k\epsilon}{l}\xi\delta U',$$
$$\delta V = \frac{1}{l}\delta V' + \frac{k\epsilon}{l}\eta\delta U',$$

and also

$$\delta W = \frac{1}{l}\delta W' + \frac{k\epsilon}{l}\zeta\delta U';$$

hence

(3) $$\sum X\delta U = \frac{1}{l}\left(kX\delta U' + Y\delta V' + Z\delta W'\right) + \frac{k\epsilon}{l}\delta U'\sum X\xi.$$

Now, in virtue of equations (2) we must have:

$$\int\sum X'\delta U'dt'\ d\tau' = \int\sum X\delta U\ dt\ d\tau = \frac{1}{l^4}\sum X\delta U\ dt'\ d\tau'.$$

By replacing $\Sigma X\delta U$ by its value (3) and by identifying, it follows:

$$X' = \frac{k}{l^5}X + \frac{k\epsilon}{l^5}\sum X\xi, \quad Y' = \frac{1}{l^5}Y, \quad Z' = \frac{1}{l^5}Z.$$

These are the equations (11) of § 1. The principle of least action leads us to the same result as the analysis of § 1.

If we turn to formulas (1), we see that Σf^2 - $\Sigma\alpha^2$ is not affected by the LORENTZ transformation, except one constant factor; it is not the case with expression $\Sigma f^2 + \Sigma\alpha^2$ which represents the energy. If we confine ourselves to the case where ϵ is sufficiently small, so that the square can be neglected so that $k = 1$, and if we also assume $l = 1$, we find:

$$\sum f'^2 = \sum f^2 + 2\epsilon\left(g\gamma - h\beta\right),$$
$$\sum \alpha'^2 = \sum \alpha^2 + 2\epsilon\left(g\gamma - h\beta\right),$$

or by addition

$$\sum f'^2 + \sum \alpha'^2 = \sum f^2 + \sum \alpha^2 + 4\epsilon\left(g\gamma - h\beta\right).$$

§ 4. — The LORENTZ group

It is important to note that the LORENTZ transformations form a group.

Indeed, if we set:

$$x' = kl(x + \epsilon t), \quad y' = ly, \quad z' = lz, \quad t' = kl(t + \epsilon x),$$

and in addition

$$x'' = k'l'(x' + \epsilon' t'), \quad y'' = l'y', \quad z'' = l'z', \quad t'' = k'l'(t' + \epsilon' x'),$$

with

$$k^{-2} = 1 - \epsilon^2, \quad k'^{-2} = 1 - \epsilon'^2$$

it follows:

$$x'' = k''l''(x + \epsilon'' t), \quad y'' = l''y, \quad z'' = l''z, \quad t'' = k''l''(t + \epsilon'' x),$$

with

$$\epsilon'' = \frac{\epsilon + \epsilon'}{1 + \epsilon\epsilon'}, \quad l'' = ll', \quad k'' = kk'(1 + \epsilon\epsilon') = \frac{1}{\sqrt{1 - \epsilon''^2}}.$$

If we take for l the value 1, and we suppose ϵ infinitely small,

$$x' = x + \delta x, \quad y' = y + \delta y, \quad z' = z + \delta z, \quad t' = t + \delta t$$

it follows:

$$\delta x = \epsilon t, \quad \delta y = \delta z = 0, \quad \delta t = \epsilon x.$$

This is the infinitesimal generator of the transformation group, which I call the transformation T_1, and which can be written in LIE's notation:

$$t\frac{d\varphi}{dx} + x\frac{d\varphi}{dt} = T_1.$$

If we assume $\epsilon = 0$ and $l = 1 + \delta l$, we find instead

$$\delta x = x\delta l, \quad \delta y = y\delta l, \quad \delta z = z\delta l, \quad \delta t = t\delta l$$

and we would have another infinitesimal transformation t_0 of the group (assuming that l and ϵ are regarded as independent variables) and we would have with LIE's notation:

$$T_0 = x\frac{d\varphi}{dx} + y\frac{d\varphi}{dy} + z\frac{d\varphi}{dz} + t\frac{d\varphi}{dt}.$$

But we could give the y- or z-axes the special role, which we gave the x-axis; thus we have two further infinitesimal transformations:

$$T_2 = t\frac{d\varphi}{dy} + y\frac{d\varphi}{dt},$$

$$T_3 = t\frac{d\varphi}{dz} + z\frac{d\varphi}{dt},$$

which also would not alter the equations of LORENTZ.

We can form combinations devised by LIE, such as

$$[T_1, T_2] = x\frac{d\varphi}{dy} - y\frac{d\varphi}{dx},$$

but it is easy to see that this transformation is equivalent to a coordinate change, the axes are rotating a very small angle around the z-axis. We should not be surprised if such a change does not alter the form of the equations of LORENTZ, obviously independent of the choice of axes.

We are thus led to consider a continuous group which we call the LORENTZ *group* and which admit as infinitesimal transformations:

1° the transformation T_0 which is permutable with all others;

2° the three transformations T_1, T_2, T_3;

3° the three rotations $[T_1, T_2]$, $[T_2, T_3]$, $[T_3, T_1]$.

Any transformation of this group can always be decomposed into a transformation of the form:

$$x' = lx, \quad y' = ly, \quad z' = lz, \quad t' = lt$$

and a linear transformation which does not change the quadratic form

$$x^2 + y^2 + z^2 - t^2$$

We can still generate our group in another way. Any transformation of the group may be regarded as a transformation of the form:

(1) $$x' = kl(x + \epsilon t), \quad y' = ly, \quad z' = lz, \quad t' = kl(t + \epsilon x)$$

preceded and followed by a suitable rotation.

But for our purposes, we should consider only a part of the transformations of this group; we must assume that *l* is a function of ε, and it is a question of choosing this function in such a way that this part of the group that I call P still forms a group.

Let's rotate the system 180° around the *y*-axis, we should find a transformation that will still belong to P. But this amounts to a sign change of *x*, *x'*, *z* and *z'*; we find:

(2) $$x' = kl(x - \epsilon t), \quad y' = ly, \quad z' = lz, \quad t' = kl(t - \epsilon x)$$

So *l* does not change when we change ε into -ε.

On the other hand, if P is a group, then the inverse substitution of (1)

(3) $$x' = \frac{k}{l}(x - \epsilon t), \quad y' = \frac{y}{l}, \quad z' = \frac{z}{l}, \quad t' = \frac{k}{l}(t - \epsilon x),$$

must also belong to P; it will therefore be identical with (2), that is to say that

$$l = \frac{1}{l}.$$

We must therefore have $l = 1$.

§ 5. — LANGEVIN waves

LANGEVIN has put the formulas that define the electromagnetic field produced by the motion of a single electron in a particularly elegant form.

Let us remember the equations

(1) $$\Box\psi = -\rho, \quad \Box F = -\rho\xi.$$

We know we can integrate by the retarded potentials and we have:

(2) $$\psi = \frac{1}{4\pi}\int \frac{\rho_1 d\tau}{r}, \quad F = \frac{1}{4\pi}\int \frac{\rho_1\xi_1 d\tau_1}{r}.$$

In these formulas we have:

$$d\tau_1 = dx_1 dy_1 dz_1, \quad r^2 = (x - x_1)^2 + (y - y_1)^2 + (z - z_1)^2,$$

whereas ρ_1 and ξ_1 are the values of ρ and ξ at the point x_1, y_1, z_1 and the instant

$$t_1 = t - r$$

x_0, y_0, z_0 being coordinates of a molecule of the electron at the instant t;

$$x_1 = x_0 + U, \quad y_1 = y_0 + V, \quad z_1 = z_0 + W$$

being its coordinates at the instant t_1;

U, V, W are functions of x_0, y_0, z_0, so that we can write:

$$dx_1 = dx_0 + \frac{dU}{dx_0}dx_0 + \frac{dU}{dy_0}dy_0 + \frac{dU}{dz_0}dz_0 + \xi_1 dt_1;$$

and if we assume t to be constant, as well as x, y and z:

$$dt_1 = +\sum \frac{x - x_1}{r}dx_1.$$

We can therefore write:

$$dx_1\left(1 + \xi_1\frac{x_1 - x}{r}\right) + dy_1\xi_1\frac{y_1 - y}{r} + dz_1\xi_1\frac{z_1 - z}{r} = dx_0\left(1 + \frac{dU}{dx_0}\right) + dy_0\frac{dU}{dy_0} + dz_0\frac{dU}{dz_0}$$

so that the other two equations can deduced by circular permutation.

We therefore have:

(3) $$d\tau_1 \left| 1 + \xi_1\frac{x_1 - x}{r}, \quad \xi_1\frac{y_1 - y}{r}, \quad \xi_1\frac{z_1 - z}{r} \right| = d\tau_0 \left| 1 + \frac{dU}{dx_0}, \quad \frac{dU}{dy_0}, \quad \frac{dU}{dz_0} \right.$$

we set

$$d\tau_0 = dx_0 dy_0 dz_0$$

Consider the determinants that appear in both sides of (3) and at the begin of the first part; if we seek to develop, we see that the terms of the 2^d and 3rd degree from ξ_1, η_1, $\zeta_1$disappear and that the determinant is equal to

$$1 + \xi_1\frac{x_1 - x}{r} + \eta_1\frac{y_1 - y}{r} + \zeta_1\frac{z_1 - z}{r} = 1 + \omega,$$

ω designates the radial component of the velocity ξ_1, η_1, ζ_1, that is to say, the component directed along the radius vector indicating from point x, y, t to point x_1, y_1, z_1.

In order to obtain the second determinant, I look at the coordinates of different molecules of the electron at instant t', which is the same for all molecules, but in such a way that for the molecule considered we have $t_1 = t'_1$. The coordinates of a molecule will then be:

$$x'_1 = x_0 + U', \quad y'_1 = y_0 + V', \quad z'_1 = z_0 + W'$$

U', V', W is what become of U, V, W, when we replace t_1 by t'_1; since t'_1 is the same for all molecules, we have:

$$dx'_1 = dx_0\left(1 + \frac{dU'}{dx_0}\right) + dy_0\frac{dU'}{dy_0} + dz_0\frac{dU'}{dz_0}$$

and therefore

$$d\tau'_1 = d\tau_0 \left| 1 + \frac{dU'}{dx_0}, \quad \frac{dU'}{dy_0}, \quad \frac{dU'}{dz_0} \right|$$

by setting

$$d\tau'_1 = dx'_1 dy'_1 dz'_1.$$

But the element of electric charge is

$$d\mu_1 = \rho_1 d\tau'_1$$

and moreover *for the molecule considered*, we have $t_1 = t'_1$, and therefore $\frac{dU'}{dx_0} = \frac{dU}{dx_0}$ etc..; we can write:

$$d\mu_1 = \rho_1 d\tau_0 \left| 1 + \frac{dU}{dx_0}, \quad \frac{dU}{dy_0}, \quad \frac{dU}{dz_0} \right|$$

so that equation (3) becomes:

$$\rho_1 d\tau_1(1 + \omega) = d\mu_1$$

and equations (2):

$$\psi = \frac{1}{4\pi}\int\frac{d\mu_1}{r(1+\omega)}, \quad F = \frac{1}{4\pi}\int\frac{\xi_1 d\mu_1}{r(1+\omega)}.$$

If we are dealing with a single electron, our integrals are reduced to a single element, provided we consider only the points x, y, x which are sufficiently remote so that r and ω have substantially the same value for all points of the electron. The potentials ψ, F, G, H depend on the position of the electron and also its velocity, because not only ξ_1, η_1, $\zeta_1$show up in the numerator of F, G, H, but the radial component ω shows up in the denominator. It is of course its position and its velocity at the instant t_1.

The partial derivatives of φ, F, G, H with respect to t, x, y, z (and therefore the electric and magnetic fields) will also depend on its acceleration. Moreover, they depend *linearly*, since the acceleration in these derivatives is introduced as a result of a single differentiation.

LANGEVIN was thus led to distinguish the electric and magnetic field terms which do not depend on the acceleration (this is what he calls the velocity wave) and those that are proportional to acceleration (that is what he calls the acceleration wave).

The calculation of these two waves is facilitated by the LORENTZ transformation. Indeed, we can apply this transformation to the system, so that the velocity of the single electron under consideration becomes zero. We will use for the x-axis the direction of the velocity before the transformation, so that, at the instant t_1,

$$\eta_1 = \zeta_1 = 0$$

and we will take ε = -ξ, so that

$$\xi_1' = \eta_1' = \zeta_1' = 0$$

We can therefore reduce the computation of the two waves to the case where the electron velocity is zero. Let's start with the velocity wave, we first note that this wave is the same as if the electron motion was uniform.

If the electron velocity is zero, then:

$$\omega = 0, \quad F = G = H = 0, \quad \psi = \frac{\mu_1}{4\pi r},$$

μ_1 is the electrical charge of the electron. The speed was reduced to zero by the LORENTZ transformation, we have now:

$$F' = G' = H' = 0, \quad \psi' = \frac{\mu_1}{4\pi r'},$$

r' is the distance from point x', y', z' at point x'_1, y'_1, z'_1, and therefore:

$$\alpha' = \beta' = \gamma' = 0,$$

$$f' = \frac{\mu_1 (x' - x_1')}{4\pi r'^3} \quad g' = \frac{\mu_1 (y' - y_1')}{4\pi r'^3}, \quad h' = \frac{\mu_1 (z' - z_1')}{4\pi r'^3}.$$

Now let us carry out the reverse LORENTZ transformation to find the true field corresponding to the velocity ε, 0, 0. We find, with reference to equations (9) and (3) of § 1:

$$(4) \quad \begin{cases} \alpha = 0, \quad \beta = \epsilon h, \quad \gamma = -\epsilon g, \\[2mm] f = \frac{\mu_1 kl^3}{4\pi r'^3} (x + \epsilon t - x_1 - \epsilon t_1), \quad g = \frac{\mu_1 kl^3}{4\pi r'^3} (y - y_1), \quad h = \frac{\mu_1 kl^3}{4\pi r'^3} (z - z \end{cases}$$

We see that the magnetic field is perpendicular to the x-axis (direction of velocity) and the electric field, and the electric field is directed to the point:

$$(5) \quad x_1 + \epsilon(t_1 - t), \quad y_1, \quad z_1.$$

If the electron continues to move in a rectilinear and uniform way with the velocity it had at the instant t_1, that is to say, with the velocity $-\varepsilon$, 0, 0, the point (5) would be the one occupied at the instant t.

Taking the acceleration wave, we can, through the LORENTZ transformation, reduce its determination to the case of zero velocity. This is the case if we imagine an electron whose oscillation amplitude is very small, but very fast, so that the displacements and velocities are much smaller, but the accelerations are finished. We thus come back to the field that has been studied in the famous work by HERTZ entitled *Die Kräfte elektrischer Schwingungen nach der Maxwell'schen Theorie*, and that for a point at great distance. In these conditions:

I° Both electric and magnetic fields are equal.

2° They are perpendicular to each other.

3° They are perpendicular to the normal of the spherical wave, that is to say to the sphere whose center is the point x_1, y_1, z_1.

I say that these three properties will remain, even when the velocity is not zero, and for this it is enough to show that they are not altered by the LORENTZ transformation.

Indeed, let A be the intensity common to both fields, let

$$(x - x_1) = r\lambda, \quad (y - y_1) = r\mu, \quad (z - z_1) = r\nu, \quad \lambda^2 + \mu^2 + \nu^2 = 1.$$

These properties expressed through the equalities

$$\begin{cases} A^2 = \sum f^2 = \sum \alpha^2, \quad \sum f\alpha = 0, \quad \sum f(x - x_1) = 0, \quad \sum \alpha(x - x_1) = 0 \\ \sum f\lambda = 0, \quad \sum \alpha\lambda = 0; \end{cases}$$

which means again that

$$\frac{b}{A}, \quad \frac{g}{A}, \quad \frac{h}{A}$$

$$\frac{\alpha}{A}, \quad \frac{\beta}{A}, \quad \frac{\gamma}{A}$$

$$\lambda, \quad \mu, \quad \nu$$

are the direction cosines of three rectangular directions, and we deduce the relations:

$$f = \beta\nu - \gamma\mu, \quad \alpha = h\mu - g\nu,$$

or

(6) $$fr = \beta(z - z_1) - \gamma(y - y_1) \quad \alpha r = h(y - y_1) - g(z - z_1),$$

with the equations that we can deduce by symmetry.

If we take the equations (3) of § 1, we find:

$$\begin{cases} x' - x_1' = kl\left[(x - x_1) + \epsilon\,(t - t_1)\right] = kl\left[(x - x_1) + \epsilon r\right], \\[2mm] y' - y_1' = l\,(y - y_1), \\[2mm] z' - z_1' = l\,(z - z_1). \end{cases}$$

(7)

We found above in § 3:

$$l^4\left(\sum f'^2 - \sum \alpha'^2\right) = \sum f^2 - \sum \alpha^2.$$

So

$$\sum f^2 = \sum \alpha^2 \quad_{\text{entrain}} \quad \sum f'^2 - \sum \alpha'^2.$$

On the other hand, from equations (9) of § 1, we get:

$$l^4 \sum f'\alpha' = \sum f\alpha,$$

This shows that

$$\sum f\alpha = 0 \quad_{\text{entrain}} \quad \sum f'\alpha' = 0.$$

I say now that

(8)
$$\sum f'\,(x' - x_1') = 0, \quad \sum \alpha'\,(x' - x_1') = 0.$$

Indeed, by virtue of equations (7) (and equations 9, § 1) the first parts of equations (8) are written respectively:

$$\frac{k}{l}\sum f\,(x - x_1) + \frac{k\epsilon}{l}\left[fr + \gamma\,(y - y_1) - \beta\,(z - z_1)\right],$$

$$\frac{k}{l}\sum \alpha\,(x - x_1) + \frac{k\epsilon}{l}\left[\alpha r - h\,(y - y_1) - g\,(z - z_1)\right].$$

They then vanish in virtue of equations $\sum f\,(x - x_1) = \sum \alpha\,(x - x_1) = 0$ and in virtue of equations (6). Yet this is precisely what was demonstrated.

We can also achieve the same result by considerations of homogeneity.

Indeed, ψ, F, G, H are functions of $x - x_1,\ y - y_1,\ z - z_1,\ \xi_1 = \dfrac{dx_1}{dt_1},\ \eta_1 = \dfrac{dy_1}{dt_1},\ \zeta_1 = \dfrac{dz_1}{dt_1}$ being homogeneous of degree -1 with respect to $x,\ y,\ z,\ t,\ x_1,\ y_1,\ z_1,\ t_1$ and their differentials.

So the derivatives of ψ, F, G, H with respect to $x,\ y,\ z,\ t$ (and hence also the two fields $f,\ g,\ h$; α, β, γ) will be homogeneous of degree -2 with respect to the same quantities, if we remember also that the relation

$$t - t_1 = r = \sqrt{\sum (x - x_1)^2}$$

is homogeneous with respect to these quantities.

But these derivatives depend on these fields of x - x₁, the velocities $\dfrac{dx_1}{dt_1}$, and the accelerations $\dfrac{d^2x_1}{dt_1^2}$; they consist of a term independent of accelerations (velocity wave) and a term linear in respect to accelerations (acceleration waves). But $\dfrac{dx_1}{dt_1}$ is homogeneous of degree 0 and $\dfrac{d^2x_1}{dt_1^2}$ is homogeneous of degree -1; hence it follows that the velocity wave is homogeneous of degree -2 with respect to x - x₁, y - y₁, z - z₁, and the acceleration wave is homogeneous of degree -1. So in a very distant point an acceleration wave is predominant and can therefore be regarded as being assimilated to the total wave. In addition, the law of homogeneity shows that the acceleration wave is similar to itself at a distance and at any point. It is therefore, at any point, similar to the total wave at a remote point. But in a distant point the disturbance can propagate as plane waves, so that the two fields should be equal, mutually perpendicular and perpendicular to the direction of propagation.

I shall refer for more details to a work by LANGEVIN in the *Journal de Physique* (Year 1905).

§ 6. — Contraction of electrons

Suppose a single electron in rectilinear and uniform motion. From what we have seen, we can, through the LORENTZ transformation, reduce the study of the field determined by the electron to the case where the electron is motionless; the LORENTZ transformation replaces the real electron in motion by an ideal electron without motion.

Let α, β, γ, f, g, h be the real field; let α', β', γ', f', g', h' be the field after the LORENTZ transformation, so the ideal field α', f' corresponds to the case where the electron is motionless; we have:

$$\alpha' = \beta' = \gamma' = 0, \quad f' = -\frac{d\psi'}{dx'}, \quad g' = -\frac{d\psi'}{dy'}, \quad h' = -\frac{d\psi'}{dz'};$$

and the actual field (in virtue of the formulas 9 of § 1):

(1)
$$\begin{cases} \alpha = 0, \quad \beta = \epsilon h, \quad \gamma = -\epsilon g \\ \\ f = l^2 f', \quad g = k l^2 g', \; h = k l^2 h'. \end{cases}$$

We now determine the total energy due to the motion of the electron, the corresponding action, and the electromagnetic momentum, in order to calculate the electromagnetic mass of the electron. For a distant point, it suffices to consider the electron as reduced to a single point; we are thus brought back to the formulas (4) of the preceding § which generally can be appropriate. But here they do not suffice, because the energy is mainly located in the ether parts nearest to the electron.

On this subject we can make several hypotheses.

According to that of ABRAHAM, the electrons are spherical and not deformable.

So when we apply the LORENTZ transformation when the real electron is spherical, the electron becomes a perfect ellipsoid. The equation of this ellipsoid is based on § 1:

$$k^2 \left(x' - \epsilon t' - \xi t' + \epsilon \xi x'\right)^2 + \left(y' - \eta k t' + \eta k \epsilon x'\right)^2$$
$$+ \left(z' - \zeta k t' + \zeta k \epsilon x'\right)^2 = l^2 r^2$$

But here we have:

$$\xi + \epsilon = \eta = \zeta = 0, \quad 1 + \epsilon \xi = 1 - \epsilon^2 = \frac{1}{k^2},$$

so that the equation of the ellipsoid becomes:

$$\frac{x'^2}{k^2} + y'^2 + z'^2 = l^2 r^2.$$

If the radius of the real electron is r, the axes of the ideal electron would therefore be:

$$klr, \ lr, \ lr.$$

In LORENTZ's hypothesis, however, the moving electrons are deformed, so that the real electron would become a ellipsoid, while the ideal electron is still always a perfect sphere of radius r, the axes of the real electron will then be:

$$\frac{r}{lk}, \quad \frac{r}{l}, \quad \frac{r}{l}.$$

We denote by

$$A = \frac{1}{2} \int f^2 d\tau$$

the *longitudinal electric energy;* by

$$B = \frac{1}{2} \int \left(g^2 + h^2 \right) d\tau$$

the *transverse electric energy;* by

$$C = \frac{1}{2} \int \left(\beta^2 + \gamma^2 \right) d\tau$$

the *transverse magnetic energy.* There is no longitudinal magnetic energy, since $\alpha = \alpha' = 0$. We denote by A', B', C' the corresponding quantities in the ideal system. We first find:

$$C' = 0, \ C = \epsilon^2 B$$

In addition, we can observe that the actual field depends only on $x = \epsilon t$, y, and x, and write:

$$d\tau = d(x + \epsilon t) dy \, dz$$
$$d\tau' = dy'dy'dz' = kl^3 d\tau$$

hence

$$A' = kl^{-1}A, \quad B' = k^{-1}l^{-1}B, \quad A = \frac{lA'}{k}, \quad B = klB'.$$

In LORENTZ's hypothesis we have B' = 2A', and A ' (being inversely proportional to the radius of the electron) is a constant independent of the velocity of the real electron; we get for the total energy:

$$A + B + C = A'lk \left(3 + \epsilon^2 \right)$$

and for the action (per unit time):

$$A + B - C = \frac{3A'l}{k}.$$

Now calculate the electromagnetic momentum; we find:

$$D = \int \left(g\gamma - h\beta \right) d\tau = -\epsilon \int \left(g^2 + h^2 \right) d\tau = -2\epsilon B = -4\epsilon kl A'.$$

But there must be some relation between the energy E = A + B + C, the action per unit time H = A + B - C, and the momentum D. The first of these relations is:

$$E = H - \epsilon \frac{dH}{d\epsilon},$$

the second is

$$\frac{dD}{d\epsilon} = -\frac{1}{\epsilon}\frac{dE}{d\epsilon};$$

hence

(2) $$D = \frac{dH}{d\epsilon}, \quad E = H - \epsilon D.$$

The second of equations (2) is always satisfied; but the first is so only if

$$l = \left(1 - \epsilon^2\right)^{\frac{1}{6}} = k^{-\frac{1}{3}},$$

that is to say if the volume of the ideal electron is equal to that of the real electron; or if the volume of the electron is constant; that's the hypothesis of LANGEVIN.

This is in contradiction with the results of § 4 and with the result obtained by LORENTZ by another way. That is the contradiction which is to be explained.

Before addressing this explanation, I note that whatever is the hypotheses we have adopted

$$H = A + B - C = \frac{l}{k}(A' + B'),$$

or, because of C' = 0,

(3) $$H = \frac{l}{k}H',$$

We can compare the result of the equation J = J' obtained in § 3.

We have indeed:

$$J = \int H \, dt, \quad J' = \int H' \, dt'.$$

We observe that the state of the system depends only on x + εt, y and z, that is to say on x', y', z', and we have:

$$t' = \frac{l}{k}t + \epsilon x'$$

(4) $$dt' = \frac{l}{k}dt.$$

By comparing equations (3) and (4) we find J = J'.

Let us consider any hypothesis, which may be either that of LORENTZ, or that of ABRAHAM, or that of LANGEVIN, or an intermediate hypothesis.

Let

$$r, \quad \theta r, \quad \theta r$$

the three axes of the real electron; that of the ideal electron will be:

$$klr, \quad \theta lr, \quad \theta lr$$

Then A' + B' is the electrostatic energy of an ellipsoid with axes *klr, θlr, θlr*.

Let us suppose that the electricity is spread on the surface of the electron as it is known from an inductor, or uniformly distributed within the electron; than this energy will be of the form:

$$A' + B' = \frac{\varphi\left(\frac{\theta}{k}\right)}{klr},$$

where φ is a known function.

The hypothesis of ABRAHAM is to assume:

$$r = const., \quad \theta = 1.$$

That of LORENTZ:

$$l = 1, \quad kr = const., \quad \theta = k.$$

That of LANGEVIN:

$$l = k^{-\frac{1}{3}}, \quad k = \theta, \quad klr = const..$$

We then find:

$$H = \frac{\varphi\left(\frac{\theta}{k}\right)}{k^2 r}.$$

ABRAHAM found, in different notation (*Göttinger Nachrichten*, 1902, p. 37)

$$H = \frac{a}{r} \frac{1 - \epsilon^2}{\epsilon} \log \frac{1 + \epsilon}{1 - \epsilon},$$

a is a constant. However, in the hypothesis of ABRAHAM, we have θ = 1; then:

(5)
$$\varphi\left(\frac{1}{k}\right) = ak^2 \frac{1 - \epsilon^2}{\epsilon} \log \frac{1 + \epsilon}{1 - \epsilon} = \frac{a}{\epsilon} \log \frac{1 + \epsilon}{1 - \epsilon},$$

which defines the function φ.

This granted, imagine that the electron is subject to a binding, so there is a relation between *r* and φ; in the hypothesis of LORENTZ this relation would be φr = const., in that ofLANGEVIN φ²r² = const. We assume in a more general way

$$r = b\theta^m$$

b is a constant; hence:

$$H = \frac{1}{bk^2}\theta^{-m}\varphi\left(\frac{\theta}{k}\right).$$

What is the shape of the electron when the velocity become -εt, *if we do not suppose the involvement of forces other than the binding forces?* Its form will be defined by the equality:

(6)
$$\frac{\partial H}{\partial \theta} = 0,$$

or

$$-m\theta^{-m-1}\varphi + \theta^{-m}k^{-1}\varphi' = 0$$

or

$$\frac{\varphi'}{\varphi} = \frac{mk}{\theta}.$$

If we want equilibrium to occur so that θ = k, it is necessary that $\frac{\theta}{k} = 1$, the logarithmic derivative of φ is equal to *m*.

If we develop $\frac{1}{k}$ and the right-hand side of (5) in powers of ε, equation (5) becomes:

$$\varphi\left(1 - \frac{\epsilon^2}{2}\right) = a\left(1 - \frac{\epsilon^2}{3}\right)$$

neglecting higher powers of ε. By differentiating, we get:

$$-\epsilon\varphi'\left(1 - \frac{\epsilon^2}{2}\right) = \frac{2}{3}\epsilon a.$$

For ε = 0, that is to say when the argument of φ is equal to 1, these equations become:

(7)
$$\varphi = a, \quad \varphi' = -\frac{2}{3}a, \quad \frac{\varphi'}{\varphi} = -\frac{2}{3}.$$

We must therefore have $m = -\frac{2}{3}$ in conformity with the hypothesis of LANGEVIN.

This result should come nearer to that which is connected to the first equation (a), and from which actually it does not differ. Indeed, suppose that every element *dτ* of the electron is subjected to a force X *dτ* parallel to the x-axis, X is the same for all elements; we will then have, in conformity with the definition of momentum:

$$\frac{dD}{dt} = \int X d\tau.$$

In addition, the principle of least action gives us:

$$\delta J = \int X \delta U \, d\tau \, dt, \quad J = \int H \, dt, \quad \delta J = \int D \delta U \, dt,$$

δU is the displacement of the center of gravity of the electron; H depends on θ and on ε if we assume that r is related to θ by the equation of binding; we have thus:

$$\delta J = \int \left(\frac{\partial H}{\partial \epsilon} \delta\epsilon + \frac{\partial H}{\partial \theta} \right) dt.$$

In addition $\delta\epsilon = -\frac{d\delta U}{dt}$; where, by integrating by parts:

$$\int D\delta\epsilon \ dt = \int D\delta U \ dt$$

or

$$\int \left(\frac{\partial H}{\partial \epsilon} \delta\epsilon + \frac{\partial H}{\partial \theta} \delta\theta \right) dt = \int D\delta\epsilon \ dt,$$

hence

$$D = \frac{\partial H}{\partial \epsilon}, \quad \frac{\partial H}{\partial \theta} = 0.$$

But the derivative $\frac{dH}{d\epsilon}$, contained in the right-hand side of equation (2), is the derivative taken by supposing θ as a function of ε, so that

$$\frac{dH}{d\epsilon} = \frac{\partial H}{\partial \epsilon} + \frac{\partial H}{\partial \theta}\frac{d\theta}{d\epsilon}.$$

Equation (2) is therefore equivalent to equation (6).

The conclusion is that if the electron is subject to a binding between its three axes, *and if no other force intervenes except the binding forces*, the shape of that electron, when it is given a uniform velocity, may be such that the ideal electron corresponds to a sphere, except the case where the binding is such that the volume is constant, in conformity with the hypothesis of LANGEVIN.

We are led in this way to pose the following problem: what additional forces, other than the binding forces, are necessary to intervene to account for the law of LORENTZ or, more generally, any law other than that of LANGEVIN?

The simplest hypothesis, and the first that we should consider, is that these additional forces are derived from a special potential depending on the three axes of the ellipsoid, and therefore on θ and on r; let F(θ, r) be the potential; in which case the action will be expressed:

$$J = \int [H + F(\theta, r)] \, dt$$

and the equilibrium conditions are written:

(8)
$$\frac{dH}{d\theta} + \frac{dF}{d\theta} = 0, \quad \frac{dH}{dr} + \frac{dF}{dr} = 0.$$

If we assume r and θ are connected by $r = b\theta^m$, we can look at r as a function of θ, consider F as depending only on θ, and retain only the first equation (8) with:

$$H = \frac{\varphi}{bk^2\theta^m}, \qquad \frac{dH}{d\theta} = \frac{-m\theta}{bk^2\theta^{m+1}} + \frac{\varphi'}{bk^3\theta^m}$$

For $k = \theta$ we need equation (8) to be satisfied; which gives, taking into account equations (7):

$$\frac{dF}{d\theta} = \frac{ma}{b\theta^{m+3}} + \frac{2}{3}\frac{a}{b\theta^{m+3}}$$

where:

$$F = \frac{-a}{b\theta^{m+2}}\frac{m+\frac{2}{3}}{m+2}$$

and in the hypothesis of LORENTZ, where $m = -1$:

$$F = \frac{a}{3b\theta}.$$

Now suppose that there is *no* connection and, considering r and θ as independent variables, retain the two equations (H); it follows:

$$H = \frac{\varphi}{k^2 r}, \qquad \frac{dH}{d\theta} = \frac{\varphi'}{k^3 r}, \qquad \frac{dH}{dr} = \frac{-\varphi}{k^2 r^2},$$

Equations (8) must be satisfied for $k = \theta$, $r = b\theta^m$; which gives:

(9)
$$\frac{dF}{dr} = \frac{a}{b^2\theta^{2m+2}}, \qquad \frac{dF}{d\theta} = \frac{2}{3}\frac{a}{b\theta^{m+3}}.$$

One way to satisfy these requirements is to pose:

(10)
$$F = Ar^\alpha\theta^\beta$$

A, α, β are constants, the equations (9) must be satisfied for $k = \theta$, $r = b\theta^m$, which gives:

$$A\alpha b^{\alpha-1}\theta^{m\alpha-m+\beta} = \frac{a}{b^2\theta^{2m+2}}, \qquad A\beta b^\alpha\theta^{m\alpha+\beta-1} = \frac{2}{3}\frac{a}{b\theta^{m+3}}.$$

By identifying we find

(11)
$$\alpha = 3\gamma, \quad \beta = 2\gamma, \quad \gamma = -\frac{m+2}{3m+2}, \quad A = \frac{a}{ab^{\alpha+1}}$$

But the volume of the ellipsoid is proportional to $r^3\theta^2$, so that the additional potential is proportional to the power γ of the volume of the electron.

In the hypothesis of LORENTZ, we have $m = 1$, $\gamma = 1$.

We thus come back to the hypothesis of LORENTZ, under the condition of adding an additional potential proportional to the volume of the electron.

The hypothesis of LANGEVIN corresponds to $\gamma = \infty$.

§ 7. — Quasi-stationary motion

It remains to see if this hypothesis of the contraction of electrons reflects the inability to demonstrate absolute motion, and I will begin by studying the quasi-stationary motion of an isolated electron, or which is subject only to the action of other distant electrons.

It is known that what is called quasi-stationary motion is the motion where the velocity changes are slow enough so that the electric and magnetic energy due to motion of the electron differ little from what they would be in uniform motion; we know also that ABRAHAM derived the transverse and longitudinal electromagnetic masses from the notion of quasi-stationary motion.

I think I should clarify. Let H be our action per unit time:

$$H = \frac{1}{2} \int \left(\sum f^2 - \sum \alpha^2 \right) d\tau,$$

where we consider for the moment only the electric and magnetic fields due to the motion of an electron. In the preceding §, by considering the motion as uniform, we regarded H as dependent from the velocity ξ, η, ζ of the electrons' center of gravity (the three components in the preceding §, had as values -ε, 0, 0) and the parameters r and θ that define the shape of the electron.

But if the motion is more uniform, H depend not only on the values of ξ, η, ζ, r, θ at the instant in question, but on values of these quantities at other instants which may differ in quantities of the same order as the time by light to travel from one point to another of the electron; in other words, H depend not only on ξ, η, ζ, r, θ, but on their derivatives of all orders with respect to time.

Well, the motion is said to be quasi-stationary when the partial derivatives of H with respect to the successive derivatives of ξ, η, ζ, r, θ are negligible compared to the partial derivatives of H with respect to the quantities ξ, η, ζ, r, θ themselves.

The equations of such a motion can be written:

(1)
$$\begin{cases} \frac{dH}{d\theta} + \frac{dF}{d\theta} = \frac{dH}{dr} + \frac{dF}{dr} = 0, \\\\ \frac{d}{dt}\frac{dH}{d\xi} = -\int X d\tau, \quad \frac{d}{dt}\frac{dH}{d\eta} = -\int Y d\tau, \quad \frac{d}{dt}\frac{dH}{d\zeta} = -\int Z d\tau. \end{cases}$$

In these equations, F has the same meaning as in the preceding §, X, Y, Z are the components of the force acting on the electron: this force is solely due to electric and magnetic fields produced by other electrons.

Note that H is independent of ξ η ζ through the combination

$$V = \sqrt{\xi^2 + \eta^2 + \zeta^2},$$

that is to say, the magnitude of the velocity; therefore we still call D the momentum:

$$\frac{dH}{d\xi} = \frac{dH}{dV}\frac{\xi}{V} = -D\frac{\xi}{V},$$

where:

$$-\frac{d}{dt}\frac{dH}{d\xi} = \frac{D}{V}\frac{d\xi}{dt} - D\frac{\xi}{V^2}\frac{dV}{dt} + \frac{dD}{dV}\frac{\xi}{V}\frac{dV}{dt},$$ (2)

$$-\frac{d}{dt}\frac{dH}{d\eta} = \frac{D}{V}\frac{d\eta}{dt} - D\frac{\eta}{V^2}\frac{dV}{dt} + \frac{dD}{dV}\frac{\eta}{V}\frac{dV}{dt},$$ (2bis)

with

$$V\frac{DV}{dt} = \sum \xi\frac{d\xi}{dt}.$$ (3)

If we take the current direction of the velocity as the x-axis, we get:

$$\xi = V, \quad \eta = \zeta = 0, \quad \frac{d\xi}{dt} = \frac{dV}{dt};$$

equations (2) and (2bis) become:

$$-\frac{d}{dt}\frac{dH}{d\xi} - \frac{dD}{dV}\frac{d\xi}{dt}, \quad -\frac{d}{dt}\frac{dH}{d\eta} = \frac{D}{V}\frac{d\eta}{dt}$$

and the last three equations (1):

$$\frac{dD}{dV}\frac{d\xi}{dt} = \int X\,d\tau, \quad \frac{D}{V}\frac{d\eta}{dt} = \int Y\,d\tau, \quad \frac{D}{V}\frac{d\zeta}{dt} = \int Z\,d\tau.$$ (4)

This is why ABRAHAM gave $\frac{dD}{dV}$ the name *longitudinal mass* and $\frac{D}{V}$ the name *transverse mass*; recall that $D = \frac{dH}{dV}$.

In the hypothesis of LORENTZ, we have:

$$D = -\frac{dH}{dV} = -\frac{\partial H}{\partial V},$$

$\frac{\partial H}{\partial V}$ represent the derivative with respect to V, after r and θ were replaced by their values as functions of V from the first two equations (1); we will also have, after the substitution,

$$H = +A\sqrt{1 - V^2}.$$

We choose units so that the constant factor A is equal to 1, and I pose $\sqrt{1 - V^2} = h$, hence:

$$H = +h, \quad D = \frac{V}{h}, \quad \frac{dD}{dV} = h^{-3}, \quad \frac{dD}{dV}\frac{1}{V^2} - \frac{D}{V^3} = h^{-3}.$$

We will pose again:

$$M = V\frac{dV}{dt} = \sum \xi\frac{d\xi}{dt}, \quad X_1 = \int X\,d\tau$$

and we find the equation for quasi-stationary motion:

(5)
$$h^{-1}\frac{d\xi}{dt} + h^{-3}\xi M = X_1.$$

Let's see what happens to these equations by the LORENTZ transformation. We will pose: 1 - ξε = μ, and we have first:

$$\mu\xi' = \xi + \epsilon, \quad \mu\eta' = \frac{\eta}{k}, \quad \mu\zeta' = \frac{\zeta}{k}$$

from which we derive easily

$$\mu h' = \frac{h}{k}.$$

We also have

$$dt' = k\,\mu\,dt$$

where:

$$\frac{d\xi'}{dt'} = \frac{d\xi}{dt}\frac{1}{k^3\mu^3}, \quad \frac{d\eta'}{dt'} = \frac{d\eta}{dt}\frac{1}{k^2\mu^2} - \frac{d\xi}{dt}\frac{\eta\epsilon}{k^2\mu^3}, \quad \frac{d\zeta'}{dt'} = \frac{d\zeta}{dt}\frac{1}{k^2\mu^2} - \frac{d\xi}{dt}\frac{\zeta\epsilon}{k^2\mu^3}$$

where again:

$$M' = \frac{d\xi}{dt}\frac{\epsilon h^2}{k^3\mu^4} + \frac{M}{k^3\mu^3}$$

and

(6)
$$h'^{-1}\frac{d\xi'}{dt'} + h'^{-3}\xi'M' = \left[h^{-1}\frac{d\xi}{dt} + h^{-3}(\xi+\epsilon)M\right]\mu^{-1},$$

(7)
$$h'^{-1}\frac{d\eta'}{dt'} + h'^{-3}\eta'M' = \left[h^{-1}\frac{d\eta}{dt} + h^{-3}\eta M\right]\mu^{-1}h^{-1}.$$

Let us return now to equations (11bis) of § 1; we can regard X_1, Y_1, Z_1 as having the same meaning as in equations (5). On the other hand, we have $l = 1$ and $\dfrac{\rho'}{\rho} = k\mu$; these equations then become:

(8)
$$\begin{cases} X_1' = \mu^{-1}\left(X_1 + \epsilon\sum X_1\xi\right), \\[2mm] Y_1' = k^{-1}\mu^{-1}Y_1. \end{cases}$$

We calculate $\Sigma X_1\xi$ using equation (5), we find:

$$\Sigma X_1\xi = h^{-3}M,$$

where:

$$\begin{cases} X_1' = \mu^{-1}\left(X_1 + \epsilon h^{-3}M\right), \\[2mm] Y_1' = k^{-1}\mu^{-1}Y_1. \end{cases}$$

Comparing equations (5) (6), (7) and (9), we finally find:

(10)

$$\begin{cases} h'^{-1}\frac{d\xi'}{dt'} + h'^{-3}\xi'M' = X_1', \\[2mm] h'^{-1}\frac{d\eta'}{dt'} + h'^{-3}\eta'M' = Y_1' \end{cases}$$

This shows that the equations of quasi-stationary motion are not altered by the LORENTZ transformation, but it still does not prove that the hypothesis of LORENTZ is the only one that leads to this result.

To establish this point, we will restrict ourselves, as LORENTZ did, to certain particular cases; it will be obviously sufficient for us to show a negative proposal.

How do we first extend the hypotheses underlying the above calculation?

1° Instead of assuming $l = 1$ in the LORENTZ transformation, we assume any l.

2° Instead of assuming that F is proportional to the volume, and hence that H is proportional to h, we assume that F is any function of θ and r, so that [after replacing θ and r with their values as functions of V, from the first two equations (1)] H is any function of V.

I note first that, assuming H = h, we must have $l = 1$; and in fact the equations (6) and (7) remain, except that the right-hand sides will be multiplied by $\frac{1}{l}$; so do equations (9), except that the right-hand sides will be multiplied by $\frac{1}{l^2}$; and finally the equations (10), except that the right-hand sides will be multiplied by $\frac{1}{l}$. If we want that the equations of motion are not altered by the LORENTZ transformation that is to say that the equations (10) only differ from equations (5) by the accentuation of the letters, it must be assumed:

$$l = 1.$$

Suppose now that we have η = ζ = 0, where ξ = V, $\frac{d\xi}{dt} = \frac{dV}{dt}$; the equations (5) take the form:

(5bis)
$$-\frac{d}{dt}\frac{dH}{d\xi} = \frac{dD}{dV}\frac{d\xi}{dt} = X_1, \qquad -\frac{d}{dt}\frac{dH}{d\eta} = \frac{D}{V}\frac{d\eta}{dt} = Y_1.$$

We can also pose:

$$\frac{dD}{dV} = f(V) = f(\xi), \qquad \frac{D}{V} = \varphi(V) = \varphi(\xi).$$

If the equations of motion are not altered by the LORENTZ transformation, we must have:

$$f(\xi)\frac{d\xi}{dt} = X_1,$$

$$\varphi(\xi)\frac{d\eta}{dt} = Y_1,$$

$$f(\xi')\frac{d\xi'}{dt'} = X_1' = l^{-2}\mu^{-1}\left(X_1 + \epsilon \sum X_1\xi\right) = l^{-2}\mu^{-1}X_1(1 + \epsilon\xi) = l^{-2}X_1,$$

$$\varphi(\xi')\frac{d\eta'}{dt'} = Y_1' = l^{-2}k^{-1}\mu^{-1}Y_1,$$

and therefore:

(11)
$$\begin{cases} f(\xi)\frac{d\xi}{dt} = l^2 f(\xi')\frac{d\xi'}{dt'}, \\[2mm] \varphi(\xi)\frac{d\eta}{dt} = l^2 k\mu\varphi(\xi')\frac{d\eta'}{dt'}. \end{cases}$$

But we have:

$$\frac{d\xi'}{dt'} = \frac{d\xi}{dt}\frac{1}{k^3\mu^3}, \quad \frac{d\eta'}{dt'} = \frac{d\eta}{dt}\frac{1}{k^2\mu^2},$$

where:

$$f(\xi') = f\left(\frac{\xi + \epsilon}{1 + \xi\epsilon}\right) = f(\xi)\frac{k^3\mu^3}{l^2},$$

$$\varphi(\xi') = \varphi\left(\frac{\xi + \epsilon}{1 + \xi\epsilon}\right) = \varphi(\xi)\frac{k\mu}{l^2};$$

whence, by eliminating l^2, we find the functional equation:

$$k^2\mu^2\frac{\varphi\left(\frac{\xi+\epsilon}{1+\xi\epsilon}\right)}{\varphi(\xi)} = \frac{f\left(\frac{\xi+\epsilon}{1+\xi\epsilon}\right)}{f(\xi)},$$

or by posing

$$\frac{\varphi(\xi)}{f(\xi)} = \Omega(\xi) = \frac{D}{V\frac{dD}{dV}},$$

that is:

$$\Omega\left(\frac{\xi + \epsilon}{1 + \xi\epsilon}\right) = \Omega(\xi)\frac{1 + \epsilon^2}{(1 + \xi\epsilon)^2},$$

an equation that must be satisfied for all values of ξ and ϵ. For $\zeta = 0$ we find:

$$\Omega(\epsilon) = \Omega(0)\left(1 - \epsilon^2\right),$$

where:

$$D = A \left(\frac{V}{\sqrt{1 - V^2}} \right)^m ,$$

A is a constant, and I set $\Omega(0) = \dfrac{1}{m}$.

We then find:

$$\varphi(\xi) = \frac{A}{\xi} \left(\frac{\xi}{\sqrt{1 - \xi^2}} \right)^m , \quad \varphi(\xi') = \frac{A\mu}{\xi + \epsilon} \left(\frac{\xi + \epsilon}{\sqrt{1 - \xi^2}\sqrt{1 - \epsilon^2}} \right)^m .$$

Now $\varphi(\xi') = \varphi(\xi)\dfrac{k\mu}{l^2}$; so we have:

$$(\xi + \epsilon)^{m-1} \left(1 - \epsilon^2 \right)^{-\frac{m}{2}} = -\xi^{m-1} \left(1 - \epsilon^2 \right)^{-\frac{1}{2}} l^{-2}.$$

As l should depend only on ϵ (since, if there are more electrons, l must be the same for all electrons whose velocities ξ may be different), this identity can take place only if we have:

$$m = 1, \ l = 1.$$

Thus LORENTZ's hypothesis is the only one consistent with the inability to demonstrate absolute motion; if we accept this impossibility, we must admit that the moving electrons contract and become ellipsoids of revolution where two axes remain constant; it must be admitted, as we have shown in the previous §, the existence of an additional potential which is proportional to the volume of the electron.

The analysis of LORENTZ is therefore fully confirmed, but we can better give us an account of the true reason of the fact which occupies us; and this reason must be sought in the considerations of § 4. *The transformations that do not alter the equations of motion must form a group, and this can take place only if l = 1.* As we do not recognize if an electron is at rest or in absolute motion, it is necessary that, when in motion, it undergoes a distortion that must be precisely that which imposes the corresponding transformation of the group.

§ 8. — Arbitrary motion

The above results apply only to quasi-stationary motion, but it is easy to extend them to the general case; it suffices to apply the principles of § 3, that is to say, the principle of least action.

For the expression of the action

$$J = \int dt \ d\tau \left(\frac{\sum f^2}{2} - \frac{\sum \alpha^2}{2} \right),$$

it is convenient to add a term representing the additional potential F of § 6; this term will obviously have the form:

$$J_1 = \int \sum (F) dt$$

where $\Sigma(F)$ represents the sum of the additional potential due to the different electrons, each of which is proportional to the volume of the corresponding electron.

I write (F) in brackets to avoid confusion with the vector F, G, H.

The total action is then $J + J_1$. We saw in § 3 that J is not altered by the LORENTZ transformation, we must show now that it is the same for J_1.

We have for one electron,

$$(F) = \omega_0 \tau$$

ω_0 being a special coefficient of the electron and τ its volume; so I can write:

$$\sum (F) = \int \omega_0 d\tau,$$

the integral has to be extended to the entire space, but so that the coefficient ω_0 is zero outside the electrons, and that within each electron it is equal to the special coefficient of that electron. Then we have:

$$J_1 = \int \omega_0 d\tau \ dt$$

and after the LORENTZ transformation:

$$J_1' = \int \omega_0' d\tau' \ dt'.$$

Now we have $\omega_0 = \omega_0'$; for if a point belong to an electron, the corresponding point after the LORENTZ transformation still belongs to the same electron. On the other hand, we found in § 3;

$$d\tau' dt' = l^4 d\tau \ dt$$

and since we now assume $l = 1$

$$d\tau' dt' = d\tau \ dt$$

We have therefore

$$J_1 = J'_1.$$ C.Q.F.D.

The theorem is thus general, it gives us at the same time a solution of the question we posed at the end of § 1: finding the complementary forces which are unaltered by theLORENTZ transformation. The additional potential (F) satisfies this condition.

So we can generalize the result announced at the end of § 1 and write:

If the inertia of electrons is exclusively of electromagnetic origin, if they are only subject to forces of electromagnetic origin, or to forces generated by the additional potential (F), no experiment can demonstrate absolute motion.

So what are these forces that create the potential (F)? They can obviously be compared to a pressure which would reign inside the electron; all occurs as if each electron were a hollow capacity subjected to a constant internal pressure (volume independent); the work of this pressure would be obviously proportional to the volume changes.

In any case, I must observe that this pressure is negative. Remember the equation (10) of § 6, according to LORENTZ's hypothesis we write:

$$F = Ar^3\theta^2;$$

equations (11) of § 6 give us:

$$A = \frac{a}{3b^4}.$$

Our pressure is equal to A, with a constant coefficient, which is indeed negative.

Now assessing the mass of the electron – I mean the "experimental mass", that is to say the mass for low velocities – we have (cf. § 6):

$$H = \frac{\varphi\left(\frac{\theta}{k}\right)}{k^2r}, \quad \theta = k, \quad \varphi = a, \quad \theta r = b;$$

hence

$$H = \frac{a}{bk} = \frac{a}{b}\sqrt{1 - V^2},$$

I can write for very small V

$$H = \frac{a}{b}\left(1 - \frac{V^2}{2}\right),$$

so that the mass, both longitudinal and transverse, will be $\frac{a}{b}$.

Now *a* is a numerical constant which shows that: *the pressure that creates our additional potential is proportional to the 4th power of the experimental mass of the electron.*

As NEWTON's law is proportional to the experimental mass, we are tempted to conclude that there is some relation between the cause that generates gravitation and the one that generates the additional potential.

§ 9. — Hypotheses on gravitation

Thus LORENTZ's theory would completely explain the impossibility to demonstrate absolute motion, if all forces are of electromagnetic origin.

But there are forces which we can not assign an electromagnetic origin, as for example gravitation. It could happen, indeed, that two systems of bodies produce equivalent electromagnetic fields, that is to say, exerting the same action on the electrified bodies and on the currents, and yet these two systems do not exercise the same gravitational action on the NEWTONian mass. The gravitational field is thus distinct from the electromagnetic field. LORENTZ was thus forced to complete his hypothesis by assuming that *forces of any origin, and in particular gravitation, are affected by translation* (or, if preferred, by the LORENTZ transformation) *the same way as electromagnetic forces*.

It is now convenient to enter into details and look more closely at this hypothesis. If we want that the NEWTONian force is affected in this way by the LORENTZ transformation, we can not accept that the force depends only on the relative position of the attracting body and of the body attracted at the instant considered. It will also depend on the velocities of the two bodies. And that's not all: it is natural to assume that the force acting at time t on the attracted body, depends on the position and velocity of this body at the same time t; but it will depend, in addition, on the position and velocity of the *attracting* body, not at time t, but *a moment earlier*, as if gravitation needs a certain time to propagate.

Consider therefore the position of the attracted body at the instant t_0 and, at this point, x_0, y_0, z_0 are the coordinates, ξ, η, ζ the components of its velocity; consider the other attracting body at the corresponding time $t_0 + t$ and, at this point, $x_0 + x$, $y_0 + y$, $z_0 + z$ are the coordinates, ξ_1, η_1, ζ_1 the components of its velocity.

We must first have a relationship

(1) $$\phi(t,\ x,\ y,\ z,\ \xi,\ \eta,\ \zeta,\ \xi_1,\ \eta_1,\ \zeta_1) = 0$$

to define the time t. This relation will define the law of propagation of the gravitational action (I do not impose on me the condition that the propagation takes place with the same speed in all directions).

Now let X_1, Y_1, Z_1 the 3 components of the action exerted at time t_0 on the body; we have to express X_1, Y_1, Z_1 as functions of

(2) $$t,\ x,\ y,\ z,\ \xi,\ \eta,\ \zeta,\ \xi_1,\ \eta_1,\ \zeta_1$$

What are the conditions to fulfill?

1° The condition (1) shall not be altered by transformations of the LORENTZ group.

2° The components X_1, Y_1, Z_1 will be affected by the LORENTZ transformations the same way as electromagnetic forces designated by the same letters, that is to say, according to equations (11[bis]) of § 1.

3° When two bodies are at rest, we will fall back to the ordinary law of attraction.

It is important to note that in the latter case, the relation (1) disappears, because time does not play any role if the two bodies are at rest.

The problem thus posed is obviously undetermined. We will thus seek to satisfy as many as possible other additional conditions:

4° Astronomical observations do not appear to show significant derogation to NEWTON's law, we will choose the solution that deviates the least of this law, for low velocities of two bodies.

5° We will endeavor to arrange that T is always negative; if indeed it is conceived that the effect of gravitation takes a certain time to be propagated, it would be more difficult to understand how this effect could depend on the position *not yet attained* by the attracting body.

There is one case where the indeterminacy of the problem disappears; it is where the two bodies are at rest *relative* to each other, that is to say that:

$$\xi = \xi_1, \quad \eta = \eta_1, \quad \zeta = \zeta_1;$$

this is the case we will consider first, assuming that these velocities are constant, so that the two bodies are drawn into a common translational motion, rectilinear and uniform.

We can assume that the axis of x has been taken parallel to the translation, so that $\eta = \zeta = 0$, and we take $\varepsilon = -\xi$.

If in these circumstances we apply the LORENTZ transformation, after the transformation the two bodies are at rest and we have:

$$\xi' = \eta' = \zeta' = 0$$

Then the components x'_0, Y'_0, Z'_0 must conform to NEWTON's law and we will have a constant factor:

(3)
$$\begin{cases} X_1' = -\frac{x'}{r'^3}, \quad Y_1' = -\frac{y'}{r'^3}, \quad Z_1' = -\frac{z'}{r'^3}, \\ \\ r'^2 = x'^2 = y'^2 + z'^2. \end{cases}$$

But we have, according to § 1:

$$x' = k(x + \epsilon t), \quad y' = y, \quad z' = z, \quad t' = k(t + \epsilon x),$$
$$\frac{\rho'}{\rho} = k(1 + \xi\epsilon) = k\left(1 - \epsilon^2\right) = \frac{1}{k}, \quad \sum X_1\xi = -X_1\epsilon,$$
$$X_1' = k\frac{\rho}{\rho'}\left(X_1 + \epsilon \sum X_1\xi\right) = k^2 X_1\left(1 - \epsilon^2\right) = X_1,$$
$$Y_1' = \frac{\rho}{\rho'}Y_1 = kY_1,$$
$$Z_1' = kZ_1.$$

We have also:

$$x + \epsilon t = x - \xi t, \quad r'^2 = k^2(x - \xi t)^2 + y^2 + z^2$$

and

(4)
$$X_1 = \frac{-k(x - \xi t)}{r'^3}, \quad Y_1 = \frac{-y}{kr'^3}, \quad Z_1 = \frac{-z}{kr'^3};$$

which can be written:

(4bis)
$$X_1 = \frac{dV}{dx}, \quad Y_1 = \frac{dV}{dy}, \quad Z_1 = \frac{dV}{dz}; \quad V = \frac{1}{kr'}.$$

It seems at first sight that the indetermination remains, since we have made no hypothesis about the value of t, that is to say about the speed of transmission; and that also x is a function of t, but it is easy to see that $x - \xi t$, y, z (which appear in our formulas) do not depend on t.

We see that if two bodies are simply in motion by a common translation, the force acting on the body is drawn normal to an ellipsoid with its center at the attracting body.

To go further we must look for the *invariants of the* LORENTZ *group*.

We know that the substitutions of this group (assuming $I = 1$) are linear substitutions which do not affect the quadratic form

$$x^2 + y^2 + z^2 - t^2.$$

Let on the other hand:

$$\xi = \frac{\delta x}{\delta t}, \quad \eta = \frac{\delta y}{\delta t}, \quad \zeta = \frac{\delta z}{\delta t};$$
$$\xi_1 = \frac{\delta_1 x}{\delta_1 t}, \quad \eta_1 = \frac{\delta_1 y}{\delta_1 t}, \quad \zeta_1 = \frac{\delta_1 z}{\delta_1 t};$$

we see that the LORENTZ transformation will cause to make δx, δy, δz and $\delta_1 x$, $\delta_1 y$, $\delta_1 z$, $\delta_1 t$ undergo the same linear substitutions as with x, y, z, t.

We regard

$$x, \qquad y, \qquad z, \qquad t\sqrt{-1},$$

$$\delta x, \qquad \delta y, \qquad \delta z, \qquad \delta t\sqrt{-1},$$

$$\delta_1 x, \qquad \delta_1 y, \qquad \delta_1 z, \qquad \delta_1 t\sqrt{-1},$$

as the coordinates of three points P, P', P" in a 4-dimensional space. We see that the LORENTZ transformation is a rotation of that space around the origin, regarded as fixed. We shall therefore have no other distinct invariants than 6 distances of the 3 points P, P', P" between them and the origin, or, if you like it better, than the 2 expressions:

$$x^2 + y^2 + z^2 - t^2, \quad x\delta x + y\delta y + z\delta z - t\delta t$$

or the 4 expressions of the same form, deduced from permuting (in an arbitrary way) the three points P, P', P".

But what we look for are the functions of 10 variables (2) that are invariants; so we must, among the combinations of our 6 invariants, seek those which depend only on these 10 variables, that is to say those that are homogeneous of degree 0 as compared to δx, δy, δz, δt, as compared to $\delta_1 x$, $\delta_1 y$, $\delta_1 z$, $\delta_1 t$. We will thus have 4 distinct invariants, which are:

$$(5) \qquad \sum x^2 - t^2, \qquad \frac{t - \sum x\xi}{\sqrt{1 - \sum \xi^2}}, \qquad \frac{t - \sum x\xi_1}{\sqrt{1 - \sum \xi_1^2}} \qquad \frac{t - \sum \xi\xi_1}{\sqrt{(1 - \sum \xi^2)(1 - \sum \xi_1^2)}}$$

Let us now consider the transformations undergone by the components of the force; resume the equations (11) of § 1, which relate not to the force X_1, Y_1, Z_1, which we consider here, but to the force X, Y, Z referred to unit volume. We pose also:

$$T = \sum X\xi;$$

we see that these equations (11) can be written as ($l = 1$):

$$(6) \qquad \begin{cases} X' = k(X + \epsilon T), & T' = k(T + \epsilon X), \\[2mm] Y' = Y, & Z' = Z; \end{cases}$$

so that X, Y, Z, T undergo the same transformation as x, y, z, t. The invariants of the group are therefore

$$\sum X^2 - T^2, \quad \sum Xx - Tt, \quad \sum X\delta x - T\delta t, \quad \sum X\delta_1 x - T\delta_1 t.$$

But this is not X, Y, Z which we need, it is X_1, Y_1, Z_1 with

$$T_1 = \sum X_1 \xi.$$

We see that

$$\frac{X_1}{X} = \frac{Y_1}{Y} = \frac{Z_1}{Z} = \frac{T_1}{T} = \frac{1}{\rho}.$$

So the LORENTZ transformations act on X_1, Y_1, Z_1, T_1 in the same manner as X, Y, Z, T, with the difference that these expressions are also multiplied by

$$\frac{\rho}{\rho'} = \frac{1}{k(1 + \xi\epsilon)} = \frac{\delta t}{\delta t'}.$$

Similarly it would act on ξ, η, ζ, 1, in the same manner as δx, δy, δz, δt, with the difference that these expressions are also multiplied by the same factor:

$$\frac{\delta t}{\delta t'} = \frac{1}{k(1 + \xi\epsilon)}.$$

Consider then X, Y, Z, $T\sqrt{-1}$ as the coordinates of a fourth point Q, then the invariants are functions of mutual distances of five points

$$0, \ P, \ P', \ P'', \ Q$$

and among these functions we must retain only those that are homogeneous of degree 0, on the one hand in relation to

$$X, \ Y, \ Z, \ T, \ \delta x, \ \delta y, \ \delta z, \ \delta t$$

(variables that can then be replaced by X_1, Y_1, Z_1, T_1, ξ, η, ζ, 1), on the other hand in relation to

$$\delta_1 x, \ \delta_1 y, \ \delta_1 z, \ 1$$

(variables that can be replaced later by ξ_1, η_1, ζ_1, 1).

Thus we find in addition to the four invariants (5), four new distinct invariants, which are:

$$(7) \quad \frac{\sum X_1^2 - T_1^2}{1 - \sum \xi^2}, \quad \frac{\sum X_1 x - T_1 t}{\sqrt{1 - \sum \xi^2}}, \quad \frac{\sum X_1 \xi_1 - T}{\sqrt{1 - \sum \xi^2}\sqrt{1 - \sum \xi_1^2}}, \quad \frac{\sum X_1 \xi - T_1}{1 - \sum \xi^2}.$$

The last invariant is always zero, according to the definition of T_1.

This granted, what are the requirements?

1° The left-hand side of relation (1), which defines the velocity of propagation must be a function of the four invariants (5)

One can obviously make a lot of hypotheses, we only look at two:

A) It may be

$$\sum x^2 - t^2 = r^2 - t^2 = 0,$$

where $t = \pm r$, and since t must be negative, $t = -r$. This means that the propagation velocity is equal to that of light. At first it seems that this hypothesis should be rejected without consideration. LAPLACE has indeed shown that this propagation is either instantaneous, or much faster than light. But LAPLACE had considered the hypothesis of finite speed of propagation, *ceteris non mutatis*; here, however, this hypothesis is complicated by many others, and it may happen that there is a more or less perfect compensation, as the applications of the LORENTZ transformation gave us already so many examples.

B) It may be

$$\frac{t - \sum x\xi_1}{\sqrt{1 - \sum \xi_1^2}} = 0, \quad t = \sum x\xi.$$

The propagation velocity is much faster than that of light, but in some cases t may be negative, which, as we have said, seems hardly acceptable. *We will add this to hypothesis (A).*

2° The four invariants (7) must be functions of the invariants (5).

3° When the two bodies are in absolute rest, X_1, Y_1, Z_1 must have the value deduced from the law of NEWTON, and when they are in relative rest, the value deduced from the equations (4).

Under the hypothesis of absolute rest, the first two invariants (7) must be reduced to

$$\sum X_1^2, \quad \sum X_1 x,$$

or by NEWTON's law at

$$\frac{1}{r^4}, \quad -\frac{1}{r}$$

secondly, in hypothesis (A), the 2nd and 3rd of the invariants (5) become:

$$\frac{-r - \sum x\xi}{\sqrt{1 - \sum \xi^2}}, \qquad \frac{-r - \sum x\xi_1}{\sqrt{1 - \sum \xi_1^2}}$$

that is to say, for absolute rest, to

$$-r, \quad -r$$

We may therefore assume for example that the first two invariants (4) are reduced to

$$\frac{(1 - \sum x\xi_1^2)^2}{(r + \sum x\xi_1)^4}, \qquad -\frac{\sqrt{1 - \sum \xi_1^2}}{r + \sum x\xi_1},$$

but other combinations are possible.

We must choose between these combinations, and secondly, in order to define X_1, Y_1, Z_1 we need a third equation. For such a choice, we must endeavor to bring us closer as much as possible to the law to NEWTON. Let's see what happens when (always making $t = -r$) we neglect the squares of the velocities ξ η etc.. The 4 invariants (5) then become:

$$0, \quad -r - \sum x\xi, \quad -r - \sum x\xi_1, \quad 1$$

and the 4 invariant (7):

$$\sum X_1^2, \quad \sum X_1(x + \xi r), \quad \sum X_1 (\xi_1 - \xi), \quad 0.$$

But to be able to compare it with the law of NEWTON, another transformation is needed; here $x_0 + x$, $y_0 + y$, $z_0 +$ z are the coordinates of the attracting body at the instant $t_0 + x$, and $r = \sqrt{\sum x^2}$; in the law of NEWTON it is necessary to consider the coordinates $x_0 + x1$, $y_0 + y_1$, $z_0 + z_1$ of the attracting body at the instant t_0, and the distance $r_1 = \sqrt{\sum x_1^2}$.

We can neglect the square of time t required for the propagation and therefore proceed as if the movement was uniform, then we have:

$$x = x_1 + \xi_1 t, \quad y = y_1 + \eta_1 t, \quad z = z_1 + \zeta_1 t, \quad r(r - r_1) = \sum x\xi_1 t$$

or, since $t = -r$,

$$x = x_1 - \xi_1 r, \quad y = y_1 - \eta_1 r, \quad z = z_1 - \zeta_1 r, \quad r = r_1 - \sum x\xi_1;$$

so that our 4 invariants (5) become:

$$0, \quad -r_1 + \sum x (\xi_1 - \xi), \quad -r_1, \quad 1$$

and our 4 invariants (7):

$$\sum X_1^2, \quad \sum X_1 [x_1 + (\xi - \xi_1) r_1], \quad \sum X_1 (\xi_1 - \xi), \quad 0.$$

In the second of these expressions I wrote r_1 instead of r, because r is multiplied by $\xi - \xi_1$ and I neglect the square of ξ.

On the other hand, NEWTON's law would us give for these 4 invariants (7)

$$\frac{1}{r_1^4}, \quad -\frac{1}{r_1} - \frac{\sum x_1 \left(\xi - \xi_1 \right)}{r_1^2}, \quad \frac{\sum x_1 \left(\xi - \xi_1 \right)}{r_1^3}, \quad 0.$$

So if we denote the 2nd and 3rd invariants (7) by A and B, and the 3 first invariants (7) by M, N, P, we will satisfy NEWTON's law up to terms of order of the square velocities, by:

(8)
$$M = \frac{1}{B^4}, \quad N = \frac{+A}{B^2}, \quad P = \frac{A-B}{B^3}.$$

This solution is not unique. Indeed, let C be the fourth invariant (5), C - 1 is of the order of the square of ξ, and it is equal to (A - B)².

So we could add to the 2ds members of each of equations (8) a term consisting of C - 1 multiplied by an arbitrary function of A, B, C, and a term of the form of (A - B)² also multiplied by a function of A, B, C.

At first sight, the solution (8) seems the most straightforward, it may nevertheless be adopted and in effect – since M, N, P are functions of X_1, Y_1, Z_1 and $T_1 = \sum X_1 \xi$ – we can draw from these three equations (8) the values of X_1, Y_1, Z_1, but in some cases these values become imaginary.

To avoid this, we will operate in another way. Let:

$$k_0 = \frac{1}{\sqrt{1 - \sum \xi^2}}, \quad k_1 = \frac{1}{\sqrt{1 - \sum \xi_1^2}}$$

This is justified by the analogy with the notation

$$k = \frac{1}{\sqrt{1 - \epsilon^2}}.$$

which appears in the substitution of LORENTZ.

In this case, and because of the condition, -r = t, the invariants (5) become:

$$0, \quad A = -k_0 \left(r + \sum x\xi \right), \quad B = -k_1 \left(r + \sum x\xi_1 \right), \quad C = k_0 k_1 \left(1 - \sum \xi\xi_1 \right).$$

On the other hand, we see that the following systems of quantities:

$$x, \quad y, \quad z, \quad -r = t$$

$$k_0 X_1, \quad k_0 Y_1, \quad k_0 Z_1, \quad k_0 T_1$$

$$k_0 \xi, \quad k_0 \eta, \quad k_0 \zeta, \quad k_0$$

$$k_1 \xi_1, \quad k_1 \eta_1, \quad k_1 \zeta_1, \quad k_1$$

undergo the *same* linear substitutions when we apply the transformations of the LORENTZ group. We are thus led to pose:

$$
\begin{cases}
X_1 = x\frac{\alpha}{k_0} + \xi\beta + \xi_1\frac{k_1}{k_0}\gamma, \\[2ex]
Y_1 = y\frac{\alpha}{k_0} + \eta\beta + \eta_1\frac{k_1}{k_0}\gamma, \\[2ex]
Z_1 = z\frac{\alpha}{k_0} + \zeta\beta + \zeta_1\frac{k_1}{k_0}\gamma, \\[2ex]
T_1 = -r\frac{\alpha}{k_0} + \beta + \frac{k_1}{k_0}\gamma,
\end{cases}
$$

(9)

It is clear that if α, β, γ are invariants, X_1, Y_1, Z_1, T_1 satisfy the basic condition, that is to say, it will undergo, by the effect of the LORENTZ transformations, a suitable linear substitution.

But for the equations (9) to be consistent, we must have:

$$\sum X_1\xi - T_1 = 0,$$

which, by replacing X_1, Y_1, Z_1, T_1 by their values (9) and multiplying by $k_0{}^2$, becomes:

(10)
$$-A\alpha - \beta - C\gamma = 0.$$

What we want is, if we neglect the square of speed of light, the squares of the velocities ξ, etc., as well as the product of accelerations by the distances as we did above, so that the values of X_1, Y_1, Z_1 remain in conformity with the law of NEWTON.

We can take:

$$\beta = 0, \quad \gamma = -\frac{A\alpha}{C}.$$

With the order of approximation adopted, we have:

$$k_0 = k_1 = 1, \quad C = 1, \quad A = -r_1 + \sum x\,(\xi_1 - \xi), \quad B = -r_1,$$
$$x = x_1 + \xi_1 t = x_1 - \xi_1 r$$

The first equation (9) becomes:

$$X_1 = \alpha\,(x - A\xi_1)$$

But if we neglect the square of ξ, we can replace $A\xi_1$ by $-r_1\xi_1$, or by $-r\xi_1$, which gives:

$$X_1 = \alpha\,(x + \xi_1 r) = \alpha x_1$$

NEWTON's law would give:

$$X_1 = -\frac{x_1}{r_1^3}.$$

We must therefore choose, for the invariant α, one that reduces to $-\frac{1}{r_1^3}$ to the order of approximation adopted, that is to say $B^{\frac{1}{3}}$. The equations (9) become:

$$(11)\quad \begin{cases} X_1 = \dfrac{x}{k_0 B^3} - \xi_1 \dfrac{k_1}{k_0} \dfrac{A}{B^3 C}, \\[2ex] Y_1 = \dfrac{y}{k_0 B^3} - \eta_1 \dfrac{k_1}{k_0} \dfrac{A}{B^3 C}, \\[2ex] Z_1 = \dfrac{z}{k_0 B^3} - \zeta_1 \dfrac{k_1}{k_0} \dfrac{A}{B^3 C}, \\[2ex] T_1 = -\dfrac{r}{k_0 B^3} - \dfrac{k_1}{k_0} \dfrac{A}{B^3 C}. \end{cases}$$

We first see that the corrected attraction is composed of two components, one parallel to the vector joining the positions of the two bodies, the other parallel to the velocity of the attracting body.

Recall that when we talk about the position or velocity of the attracting body, it is its position or its velocity when the gravitational wave leaves; for the body attracted, on the contrary, it is its position or its velocity when the gravitational wave reaches it, the wave is assumed to propagate with the speed of light.

I think it would be premature to push further discussion of these formulas, I will confine myself to a few remarks.

1° The solutions (11) are not unique; we can indeed replace $\dfrac{1}{B^3}$ which enters in the factor everywhere, by

$$\frac{1}{B^3} + (C - 1) f_1(A, B, C) + (A - B)^2 f_2(A, B, C)$$

f_1 and f_2 are arbitrary functions of A, B, C; or we are taking β no longer as zero but adding arbitrary complementary terms to α β γ, provided they satisfy the condition (10) and are of the 2nd order with regard to ξ as far as α is concerned, and of the 1st order as far as β and γ are concerned.

2° The first equation (11) can be written:

$$(11^{bis})\quad X_1 \frac{k_1}{B^3 C} \left[x \left(1 - \sum \xi \xi_1 \right) + \xi_1 \left(r + \sum x\xi \right) \right]$$

and the quantity in brackets can, itself, written as:

$$(12)\quad (x + r\xi_1) + \eta\left(\xi_1 y - x\eta_1\right) + \zeta\left(\xi_1 z - x\zeta_1\right),$$

so that the total force can be divided into three components corresponding to the three brackets of expression (12); the first component has a vague analogy with the mechanical force due to the electric field, the other two with mechanical forces due to a magnetic field; to complete the analogy I can, under the first point, replace $\dfrac{1}{B^3}$ by $\dfrac{C}{B^3}$ in equations (11), so that X_1, Y_1, Z_1 only depend linearly on the velocity ξ, η, ζ of the attracted body, since C has disappeared from the denominator of (11bis).

We pose then:

$$(13) \quad \begin{cases} k_1(x + r\xi_1) = \lambda, \quad k_1(y + r\eta_1) = \mu, \quad k_1(z + r\zeta_1) = \nu, \\[2mm] k_1(\eta_1 z - \zeta_1 y) = \lambda', \quad k_1(\zeta_1 x - \xi_1 z) = \mu', \quad k_1(\xi_1 y - x\eta_1) = \nu'; \end{cases}$$

it follows that C had disappeared from the denominator of (11a):

$$(14) \quad \begin{cases} X_1 = \dfrac{\lambda}{B^3} + \dfrac{\eta\nu' - \zeta\mu'}{B^3}, \\[3mm] Y_1 = \dfrac{\mu}{B^3} + \dfrac{\zeta\lambda' - \xi\nu'}{B^3}, \\[3mm] Z_1 = \dfrac{\nu}{B^3} + \dfrac{\xi\mu' - \eta\lambda'}{B^3}, \end{cases}$$

and there will also:

$$(15) \quad B^2 = \sum \lambda^2 - \sum \lambda'^2.$$

Then λ, μ, v or $\dfrac{\lambda}{B^3}$, $\dfrac{\mu}{B^3}$, $\dfrac{\nu}{B^3}$ is a kind of electric field, while λ', μ', v' or rather $\dfrac{\lambda'}{B^3}$, $\dfrac{\mu'}{B^3}$, $\dfrac{\nu'}{B^3}$ is a kind of magnetic field.

3° The postulate of relativity would require us to adopt solution (11) or solution (14) or any solution that would inferred by using the first remark; but the first question that arises is whether they are compatible with astronomical observations; the discrepancy with NEWTON's law is of the order ξ^2, that is to say, 10000 times smaller when it were of order ξ, that is to say, if the propagation happens with the speed of light, *ceteris non mutatis*; it is permissible to hope that it will not be too great. But only a thorough discussion will be able to teach it to us.

Paris, July 1905.

H. POINCARÉ

----

1. LANGEVIN was preceded by M. BUCHERER from Bonn, who had put forward the same theory before. (*See*: BUCHERER, *Mathematische Einführung in die Elektronentheorie*; August 1904. Teubner, Leipzig).

The End of Matter (1906)
by Henri Poincaré, translated from French by Wikisource

The End of Matter.[1]

One of the most surprising discoveries made by the physicists in recent years, amounts in the claim that matter doesn't exist. We simultaneously add, that this discovery is not definitely established. The essential property of matter is its mass and its inertia. Mass remains constant everywhere and always, it even remains when a chemical transformation changes all observable properties of matter, and has apparently created an entirely new body. Thus if one is able to show, that matter carries mass like a foreign jewelry, that this mass (always considered as constant) can also suffer variations: then one surely has the right to say that there is no matter. But this is precisely what is announced.

The velocities observable up to now, are only small, because the celestial bodies, although much faster than automobiles, hardly make 60 or 100 kilometers per second; yet, light is ca. 3000 times faster, thus we aren't dealing with moving matter, but with a disturbance of equilibrium which propagates in a relatively unmovable substance, like a wave at the surface of the sea. All experiments conducted at these small velocities, have always confirmed the constancy of mass, and nobody has asked himself the question, whether this law is also valid at higher velocities.

The speed record of mercury (the fastest planet) was broken by those infinitely small bodies: I'm talking about the corpuscles by whose motions the cathode rays and the radium rays arise. It's known that these emanations were caused by a veritable bombardment of molecules. The projectiles ejected at this occasion, are charged with negative electricity; one can convince oneself of this fact by collecting this electricity with a suitable apparatus. In consequence of this charge, they will be deflected by a magnetic or an electric field, and by measuring these deflections one can determine their velocity and the ratio of their charge to their mass.

On one hand such measurement have taught us that their velocity is enormously great, by achieving approximately a tenth or a third of the speed of light, and thus being a thousand times greater than the velocity of the planets; on the other hand they have taught us that their charge with respect to their mass is considerable. Any moving corpuscle thus represents an electric current. Now, it's known that electric currents show a special kind of inertia, which is denoted as *self-induction*. Once created, a current has the endeavor to conserve itself; from that it comes that one notices a jumping of a spark, when one cuts the conductor (which is traversed by a current) and thus current is interrupted. The current endeavors to remain its intensity in the same way as a moving body endeavors to remain its velocity. Also our cathode-corpuscle will have a certain resistance with respect to the influences which can change its velocity: first by its actual inertia, second by its self-induction – the latter is the case because every change of its velocity would be connected with a simultaneous change of the corresponding current. The *electrons* – the name of the corpuscles – thus would have two kinds of inertia: the mechanical inertia and the electromagnetic inertia.

The works of the theoretician *Abraham* and the experimentalist *Kaufmann* were aimed to specify these two kinds of inertia more closely. For this purpose they had to made a hypothesis; they assumed that all negative electrons are identical with each other, that they all have the same essentially constant charge, and that the differences which exist among them are only caused by their different velocities. When the velocity is changing then their real, *i.e.* their mechanical mass, remains constant; this is so to speak the definition of the latter. However, the electromagnetic mass which causes the apparent mass, increases with velocity by a certain law. Therefore a certain relation must exist between the velocity and the ratio of mass to charge; as it was already said above, one can calculate both quantities by observing the deflections which were suffered by the rays under the influence of a magnet or an electric field; the study of these relations allows to separately determine the amount of both inertias. This result is totally surprising: *the real mass is equal to zero*. This conclusion is based, however, on the hypothesis mentioned before, but the agreement between the theoretical and the experimental curve is at least great enough, to make this hypothesis plausible.

Consequently, these negative electrons have no actual mass; if they still appear to be equipped with inertia, then this due to the fact that their velocity cannot be changed without a simultaneous disturbance of the luminiferous aether. Their apparent inertia is only borrowed, it doesn't belong to them, but to the aether. Yet, matter doesn't entirely consists of negative electrons; one can rather assume that there also exists real matter which possesses a certain inertia. There are rays also due to rain of projectiles, yet those projectiles carry positive charges with them: the canal-rays of Goldstein and the rays of radium belong to them; do those positive electrons also have no mass? This cannot be said, because they are much heavier than the negative electrons and are moving much slower. Here, two hypotheses are possible: either the electrons are heavier because they have a particular mechanical inertia besides the electromagnetic inertia from the aether, thus they would form the actual matter; or they are without mass as well, and only appear to be heavier because they are much smaller. I intentionally say "much smaller", even hough this may appear paradoxical; by this assumption, the corpuscles would only represent holes in the aether, and only the aether alone would actually exist and be endowed with inertia.

So far, the existence of matter wouldn't be endangered a lot; we still can decide ourselves for the first hypothesis, we even can assume that there also exist other atoms besides the positive and negative electrons. However, this last refuge was also taken away from us by the recent investigations of *Lorentz*. We are carried by Earth in its extremely fast motion; should the optical and electrical phenomena not at all be influenced by this translation? For a long time, one has believed in such an influence, one has assumed that it would be possible to demonstrate (depending on the orientation of the apparatus with respect to Earth's motion) differences in the observations. This expectation was in vain, even the most precise measurements never demonstrated anything like that. So by that, the experiments justified a conviction common to all physicists: if it were possible to demonstrate any influence, then one would be in the position, not only to determine the relative motion of Earth around the sun, but even its absolute motion in the aether. For many persons it will be hard to believe, that something different than relative motion can ever be demonstrated by experiment; they rather stick to the conclusion that matter has no mass.

Thus we weren't too much surprised by the negative results; although they contradicted the predominant theories, they satisfied a certain deeper instinct, which is older and stronger than all theories. There remained nothing left than to change the theories, so that they can be brought into agreement with this fact again. For this purpose, *Fitzgerald* made a surprising hypothesis: according to him, alls bodies shall suffer a contraction of about a hundred-millionth in the direction of Earth's motion. A perfect sphere consequently changes into an oblate ellipsoid, and if one sets it into rotation, than it is deformed so that the minor axis of the ellipsoid always remains parallel to the direction of Earth's velocity. Since the measuring instruments are deformed in the same ways as the observed objects, the deformation cannot be observed, unless one could determine the time which is required by light to traverse the object along its length.

This hypotheses accounts for the observed facts. But one cannot be satisfied by that; sometimes one will make observations which are still more precise: will the results be positive? will they give us the means to determine the absolute motion of Earth? *Lorentz* doesn't believe this; he thinks that such a determination is impossible also in the future; the agreeing instinct of all physicists, the failure of all experiments up to now, sufficiently justifies his view. Thus let us consider this impossibility as a general law of nature, let us take it as a postulate. Which consequences follow from that? This question was further investigated by *Lorentz*; he found that all atoms as well as all positive and negative electrons have an inertia, which (for all of them) varies with velocity by the same laws. Any material atom would therefore be composed by small and heavy positive electrons, and when the observable matter doesn't appear to us as electric, then this is caused by the fact that both kinds of electrons are present in approximately the same amount. They all have no measures, and their inertia is borrowed from the aether. In this system there is no actual matter, there are only holes in the aether.

According to *Langevin*, matter is liquefied aether that has changed its properties; when matter is moving, then it is not the liquefied mass that travels in the aether, but the liquefaction still affects new parts of the aether; and the aether that is located behind moving matter (and which was liquefied before) has returned to its earlier rigid state. Therefore, moving matter doesn't remain identical with itself.

This was the state of facts until recently; but now *Kaufmann* arrives with new experiments. The negative electron, whose velocity is extraordinarily great, must also suffer the contraction assumed by *Fitzgerald*, and by that the relation between velocity and mass would be modified; however, this wasn't confirmed by the recent experiments; thus the whole building seems to break down, and matter seems to keep its justification to exist. Besides, it is only about very small quantities in these experiments, and therefore a definite decision would still be too early today.

1. See the book by Gustave Le Bon: l'Evolution de la Matière.

The New Mechanics (1908)
by Henri Poincaré, translated by George Bruce Halsted

CHAPTER I

Mechanics and Radium

I

Introduction

THE general principles of Dynamics, which have, since Newton, served as foundation for physical science, and which appeared immovable, are they on the point of being abandoned or at least profoundly modified? This is what many people have been asking themselves for some years. According to them, the discovery of radium has overturned the scientific dogmas we believed the most solid: on the one hand, the impossibility of the transmutation of metals; on the other hand, the fundamental postulates of mechanics.

Perhaps one is too hasty in considering these novelties as finally established, and breaking our idols of yesterday; perhaps it would be proper, before taking sides, to await experiments more numerous and more convincing. None the less is it necessary, from to-day, to know the new doctrines and the arguments, already very weighty, upon which they rest.

In few words let us first recall in what those principles consist:

A. The motion of a material point isolated and apart from all exterior force is straight and uniform; this is the principle of inertia: without force no acceleration;

B. The acceleration of a moving point has the same direction as the resultant of all the forces to which it is subjected; it is equal to the quotient of this resultant by a coefficient called *mass* of the moving point.

The mass of a moving point, so defined, is a constant; it does not depend upon the velocity acquired by this point; it is the same whether the force, being parallel to this velocity, tends only to accelerate or to retard the motion of the point, or whether, on the contrary, being perpendicular to this velocity, it tends to make this motion deviate toward the right, or the left, that is to say to *curve* the trajectory;

C. All the forces affecting a material point come from the action of other material points; they depend only upon the *relative* positions and velocities of these different material points.

Combining the two principles B and C, we reach the *principle of relative motion*, in virtue of which the laws of the motion of a system are the same whether we refer this system to fixed axes, or to moving axes animated by a straight and uniform motion of translation, so that it is impossible to distinguish absolute motion from a relative motion with reference to such moving axes;

D. If a material point A acts upon another material point B, the body B reacts upon A, and these two actions are two equal and directly opposite forces. This is *the principle of the equality of action and reaction*, or, more briefly, the *principle of reaction*.

Astronomic observations and the most ordinary physical phenomena seem to have given of these principles a confirmation complete, constant and very precise. This is true, it is now said, but it is because we have never operated with any but very small velocities; Mercury, for example, the fastest of the planets, goes scarcely 100 kilometers a second. Would this planet act the same if it went a thousand times faster? We see there is yet no need to worry; whatever may be the progress of automobilism, it will be long before we must give up applying to our machines the classic principles of dynamics.

How then have we come to make actual speeds a thousand times greater than that of Mercury, equal, for instance, to a tenth or a third of the velocity of light, or approaching still more closely to that velocity? It is by aid of the cathode rays and the rays from radium.

We know that radium emits three kinds of rays, designated by the three Greek letters α, β, γ; in what follows, unless the contrary be expressly stated, it will always be a question of the β rays, which are analogous to the cathode rays.

After the discovery of the cathode rays two theories appeared: Crookes attributed the phenomena to a veritable molecular bombardment; Hertz, to special undulations of the ether. This was a renewal of the debate which divided physicists a century ago about light; Crookes took up the emission theory, abandoned for light; Hertz held to the undulatory theory. The facts seem to decide in favor of Crookes.

It has been recognized, in the first place, that the cathode rays carry with them a negative electric charge; they are deviated by a magnetic field and by an electric field; and these deviations are precisely such as these same fields would produce upon projectiles animated by a very high velocity and strongly charged with electricity. These two deviations depend upon two quantities: one the velocity, the other the relation of the electric charge of the projectile to its mass; we cannot know the absolute value of this mass, nor that of the charge, but only their relation; in fact, it is clear that if we double at the same time the charge and the mass, without changing the velocity, we shall double the force which tends to deviate the projectile, but, as its mass is also doubled, the acceleration and deviation observable will not be changed. The observation of the two deviations will give us therefore two equations to determine these two unknowns. We find a velocity of from 10,000 to 30,000 kilometers a second; as to the ratio of the charge to the mass, it is very great. We may compare it to the corresponding ratio in regard to the hydrogen ion in electrolysis; we then find that a cathodic projectile carries about a thousand times more electricity than an equal mass of hydrogen would carry in an electrolyte.

To confirm these views, we need a direct measurement of this velocity to compare with the velocity so calculated. Old experiments of J. J. Thomson had given results more than a hundred times too small; but they were exposed to certain causes of error. The question was taken up again by Wiechert in an arrangement where the Hertzian oscillations were utilized; results were found agreeing with the theory, at least as to order of magnitude; it would be of great interest to repeat these experiments. However that may be, the theory of undulations appears powerless to account for this complex of facts.

The same calculations made with reference to the β rays of radium have given velocities still greater: 100,000 or 200,000 kilometers or more yet. These velocities greatly surpass all those we know. It is true that light has long been known to go 300,000 kilometers a second; but it is not a carrying of matter, while, if we adopt the emission theory for the cathode rays, there would be material molecules really impelled at the velocities in question, and it is proper to investigate whether the ordinary laws of mechanics are still applicable to them.

II

Mass Longitudinal and Mass Transversal

We know that electric currents produce the phenomena of induction, in particular *self-induction*.

When a current increases, there develops an electromotive force of self-induction which tends to oppose the current; on the contrary, when the current decreases, the electromotive force of self-induction tends to maintain the current. The self-induction therefore opposes every variation of the intensity of the current, just as in mechanics the inertia of a body opposes every variation of its velocity.

Self-induction is a veritable inertia. Everything happens as if the current could not establish itself without putting in motion the surrounding ether and as if the inertia of this ether tended, in consequence, to keep constant the intensity of this current. It would be requisite to overcome this inertia to establish the current, it would be necessary to overcome it again to make the current cease.

A cathode ray, which is a rain of projectiles charged with negative electricity, may be likened to a current; doubtless this current differs, at first sight at least, from the currents of ordinary conduction, where the matter does not move and where the electricity circulates through the matter. This is a *current of convection*, where the electricity, attached to a material vehicle, is carried along by the motion of this vehicle. But Rowland has proved that currents of convection produce the same magnetic effects as currents of conduction; they should produce also the same effects of induction. First, if this were not so, the principle of the conservation of energy would be violated; besides, Crémieu and Pender have employed a method putting in evidence *directly* these effects of induction.

If the velocity of a cathode corpuscle varies, the intensity of the corresponding current will likewise vary; and there will develop effects of self-induction which will tend to oppose this variation. These corpuscles should therefore possess a double inertia: first their own proper inertia, and then the apparent inertia, due to self-induction, which produces the same effects. They will therefore have a total apparent mass, composed of their real mass and of a fictitious mass of electromagnetic origin. Calculation shows that this fictitious mass varies with the velocity, and that the force of inertia of self-induction is not the same when the velocity of the projectile accelerates or slackens, or when it is deviated; therefore so it is with the force of the total apparent inertia.

The total apparent mass is therefore not the same when the real force applied to the corpuscle is parallel to its velocity and tends to accelerate the motion as when it is perpendicular to this velocity and tends to make the direction vary. It is necessary therefore to distinguish the *total longitudinal mass* from the *total transversal mass*. These two total masses depend, moreover, upon the velocity. This follows from the theoretical work of Abraham.

In the measurements of which we speak in the preceding section, what is it we determine in measuring the two deviations? It is the velocity on the one hand, and on the other hand the ratio of the charge to the *total transversal mass*. How, under these conditions, can we make out in this total mass the part of the real mass and that of the fictitious electromagnetic mass? If we had only the cathode rays properly so called, it could not be dreamed of; but happily we have the rays of radium which, as we have seen, are notably swifter. These rays are not all identical and do not behave in the same way under the action of an electric field and a magnetic field. It is found that the electric deviation is a function of the magnetic deviation, and we are able, by receiving on a sensitive plate radium rays which have been subjected to the action of the two fields, to photograph the curve which represents the relation between these two deviations. This is what Kaufmann has done, deducing from it the relation between the velocity and the ratio of the charge to the total apparent mass, a ratio we shall call ε.

One might suppose there are several species of rays, each characterized by a fixed velocity, by a fixed charge and by a fixed mass. But this hypothesis is improbable; why, in fact, would all the corpuscles of the same mass take always the same velocity? It is more natural to suppose that the charge as well as the real mass are the same for all the projectiles, and that these differ only by their velocity. If the ratio ε is a function of the velocity, this is not because the real mass varies with this velocity; but, since the fictitious electromagnetic mass depends upon this velocity, the total apparent mass, alone observable, must depend upon it, though the real mass does not depend upon it and may be constant.

The calculations of Abraham let us know the law according to which the *fictitious* mass varies as a function of the velocity; Kaufmann's experiment lets us know the law of variation of the *total* mass.

The comparison of these two laws will enable us therefore to determine the ratio of the *real* mass to the total mass.

Such is the method Kaufmann used to determine this ratio. The result is highly surprising: *the real mass is naught.*

This has led to conceptions wholly unexpected. What had only been proved for cathode corpuscles was extended to all bodies. What we call mass would be only semblance; all inertia would be of electromagnetic origin. But then mass would no longer be constant, it would augment with the velocity; sensibly constant for velocities up to 1,000 kilometers a second, it then would increase and would become infinite for the velocity of light. The transversal mass would no longer be equal to the longitudinal: they would only be nearly equal if the velocity is not too great. The principle *B* of mechanics would no longer be true.

III

The Canal Rays

At the point where we now are, this conclusion might seem premature. Can one apply to all matter what has been proved only for such light corpuscles, which are a mere emanation of matter and perhaps not true matter? But before entering upon this question, a word must be said of another sort of rays. I refer to the *canal rays*, the *Kanalstrahlen* of Goldstein.

The cathode, together with the cathode rays charged with negative electricity, emits canal rays charged with positive electricity. In general, these canal rays not being repelled by the cathode, are confined to the immediate neighborhood of this cathode, where they constitute the 'chamois cushion,' not very easy to perceive; but, if the cathode is pierced with holes and if it almost completely blocks up the tube, the canal rays spread *back* of the cathode, in the direction opposite to that of the cathode rays, and it becomes possible to study them. It is thus that it has been possible to show their positive charge and to show that the magnetic and electric deviations still exist, as for the cathode rays, but are much feebler.

Radium likewise emits rays analogous to the canal rays, and relatively very absorbable, called α rays.

We can, as for the cathode rays, measure the two deviations and thence deduce the velocity and the ratio ε. The results are less constant than for the cathode rays, but the velocity is less, as well as the ratio ε; the positive corpuscles are less charged than the negative; or if, which is more natural, we suppose the charges equal and of opposite sign, the positive corpuscles are much the larger. These corpuscles, charged the ones positively, the others negatively, have been called *electrons.*

IV

The Theory of Lorentz

But the electrons do not merely show us their existence in these rays where they are endowed with enormous velocities. We shall see them in very different roles, and it is they that account for the principal phenomena of optics and electricity. The brilliant synthesis about to be noticed is due to Lorentz.

Matter is formed solely of electrons carrying enormous charges, and, if it seems to us neutral, this is because the charges of opposite sign of these electrons compensate each other. We may imagine, for example, a sort of solar system formed of a great positive electron, around which gravitate numerous little planets, the negative electrons, attracted by the electricity of opposite name which charges the central electron. The negative charges of these planets would balance the positive charge of this sun, so that the algebraic sum of all these charges would be naught.

All these electrons swim in the ether. The ether is everywhere identically the same, and perturbations in it are propagated according to the same laws as light or the Hertzian oscillations *in vacuo*. There is nothing but electrons and ether. When a luminous wave enters a part of the ether where electrons are numerous, these electrons are put in motion under the influence of the perturbation of the ether, and they then react upon the ether. So would be explained refraction, dispersion, double refraction and absorption. Just so, if for any cause an electron be put in motion, it would trouble the ether around it and would give rise to luminous waves, and this would explain the emission of light by incandescent bodies.

In certain bodies, the metals for example, we should have fixed electrons, between which would circulate moving electrons enjoying perfect liberty, save that of going out from the metallic body and breaking the surface which separates it from the exterior void or from the air, or from any other non-metallic body.

These movable electrons behave then, within the metallic body, as do, according to the kinetic theory of gases, the molecules of a gas within the vase where this gas is confined. But, under the influence of a difference of potential, the negative movable electrons would tend to go all to one side, and the positive movable electrons to the other. This is what would produce electric currents, and *this is why these bodies would he conductors*. On the other hand, the velocities of our electrons would be the greater the higher the temperature, if we accept the assimilation with the kinetic theory of gases. When one of these movable electrons encounters the surface of the metallic body, whose boundary it can not pass, it is reflected like a billiard ball which has hit the cushion, and its velocity undergoes a sudden change of direction. But when an electron changes direction, as we shall see further on, it becomes the source of a luminous wave, and this is why hot metals are incandescent.

In other bodies, the dielectrics and the transparent bodies, the movable electrons enjoy much less freedom. They remain as if attached to fixed electrons which attract them. The farther they go away from them the greater becomes this attraction and tends to pull them back. They therefore can make only small excursions; they can no longer circulate, but only oscillate about their mean position. This is why these bodies would not be conductors; moreover they would most often be transparent, and they would be refractive, since the luminous vibrations would be communicated to the movable electrons, susceptible of oscillation, and thence a perturbation would result.

I can not here give the details of the calculations; I confine myself to saying that this theory accounts for all the known facts, and has predicted new ones, such as the Zeeman effect.

V

Mechanical Consequences

We now may face two hypotheses :

1° The positive electrons have a real mass, much greater than their fictitious electromagnetic mass;

the negative electrons alone lack real mass. We might even suppose that apart from electrons of the two signs, there are neutral atoms which have only their real mass. In this case, mechanics is not affected; there is no need of touching its laws; the real mass is constant; simply, motions are deranged by the effects of self-induction, as has always been known; moreover, these perturbations are almost negligible, except for the negative electrons which, not having real mass, are not true matter;

2° But there is another point of view; we may suppose there are no neutral atoms, and the positive electrons lack real mass just as the negative electrons. But then, real mass vanishing, either the word *mass* will no longer have any meaning, or else it must designate the fictitious electromagnetic mass; in this case, mass will no longer be constant, the transversal mass will no longer be equal to the longitudinal, the principles of mechanics will be overthrown.

First a word of explanation. We have said that, for the same charge, the *total* mass of a positive electron is much greater than that of a negative. And then it is natural to think that this difference is explained by the positive electron having, besides its fictitious mass, a considerable real mass; which takes us back to the first hypothesis. But we may just as well suppose that the real mass is null for these as for the others, but that the fictitious mass of the positive electron is much the greater since this electron is much the smaller. I say advisedly: much the smaller. And, in fact, in this hypothesis inertia is exclusively electromagnetic in origin; it reduces itself to the inertia of the ether; the electrons are no longer anything by themselves; they are solely holes in the ether and around which the ether moves; the smaller these holes are, the more will there be of ether, the greater, consequently, will be the inertia of the ether.

How shall we decide between these two hypotheses? By operating upon the canal rays as Kaufmann did upon the β rays? This is impossible; the velocity of these rays is much too slight. Should each therefore decide according to his temperament, the conservatives going to one side and the lovers of the new to the other? Perhaps, but, to fully understand the arguments of the innovators, other considerations must come in.

CHAPTER II

Mechanics and Optics

Aberration

You know in what the phenomenon of aberration, discovered by Bradley, consists. The light issuing from a star takes a certain time to go through a telescope; daring this time, the telescope, carried along by the motion of the earth, is displaced. If therefore the telescope were pointed in the *true* direction of the star, the image would be formed at the point occupied by the crossing of the threads of the network when the light has reached the objective; and this crossing would no longer be at this same point when the light reached the plane of the network. We would therefore be led to mis-point the telescope to bring the image upon the crossing of the threads. Thence results that the astronomer will not point the telescope in the direction of the absolute velocity of the light, that is to say toward the true position of the star, but just in the direction of the relative velocity of the light with reference to the earth, that is to say toward what is called the apparent position of the star.

The velocity of light is known; we might therefore suppose that we have the means of calculating the *absolute* velocity of the earth. (I shall soon explain my use here of the word absolute.) Nothing of the sort; we indeed know the apparent position of the star we observe ; but we do not know its true position; we know the velocity of the light only in magnitude and not in direction.

If therefore the absolute velocity of the earth were straight and uniform, we should never have suspected the phenomenon of aberration; but it is variable; it is composed of two parts: the velocity of the solar system, which is straight and uniform; the velocity of the earth with reference to the sun, which is variable. If the velocity of the solar system, that is to say if the constant part existed alone, the observed direction would be invariable. This position that one would thus observe is called the mean apparent position of the star.

Taking account now at the same time of the two parts of the velocity of the earth, we shall have the actual apparent position, which describes a little ellipse around the mean apparent position, and it is this ellipse that we observe.

Neglecting very small quantities, we shall see that the dimensions of this ellipse depend only upon the ratio of the velocity of the earth with reference to the sun to the velocity of light, so that the relative velocity of the earth with regard to the sun has alone come in.

But wait! This result is not exact, it is only approximate; let us push the approximation a little farther. The dimensions of the ellipse will depend then upon the absolute velocity of the earth. Let us compare the major axes of the ellipse for the different stars: we shall have, theoretically at least, the means of determining this absolute velocity.

That would be perhaps less shocking than it at first seems; it is a question, in fact, not of the velocity with reference to an absolute void, but of the velocity with regard to the ether, which is taken *by definition* as being absolutely at rest.

Besides, this method is purely theoretical. In fact, the aberration is very small; the possible variations of the ellipse of aberration are much smaller yet, and, if we consider the aberration as of the first order, they should therefore be regarded as of the second order: about a millionth of a second; they are absolutely inappreciable for our instruments. We shall finally see, further on, why the preceding theory should be rejected, and why we could not determine this absolute velocity even if our instruments were ten thousand times more precise!

One might imagine some other means, and in fact, so one has. The velocity of light is not the same in water as in air; could we not compare the two apparent positions of a star seen through a telescope first full of air, then full of water? The results have been negative; the apparent laws of reflection and refraction are not altered by the motion of the earth. This phenomenon is capable of two explanations :

1° It might be supposed that the ether is not at rest, but that it is carried along by the body in motion. It would then not be astonishing that the phenomena of refraction are not altered by the motion of the earth, since all, prisms, telescopes and ether, are carried along together in the same translation. As to the aberration itself, it would be explained by a sort of refraction happening at the surface of separation of the ether at rest in the interstellar spaces and the ether carried along by the motion of the earth. It is upon this hypothesis (bodily carrying along of the ether) that is founded the *theory of Hertz* on the electrodynamics of moving bodies.

2° Fresnel, on the contrary, supposes that the ether is at absolute rest in the void, at rest almost absolute in the air, whatever be the velocity of this air, and that it is partially carried along by refractive media. Lorentz has given to this theory a more satisfactory form. For him, the ether is at rest, only the electrons are in motion; in the void, where it is only a question of the ether, in the air, where this is almost the case, the carrying along is null or almost null; in refractive media, where perturbation is produced at the same time by vibrations of the ether and those of electrons put in swing by the agitation of the ether, the undulations are *partially* carried along.

To decide between the two hypotheses, we have Fizeau's experiment, comparing by measurements of the fringes of interference, the velocity of light in air at rest or in motion. These experiments have confirmed Fresnel's hypothesis of partial carrying along. They have been repeated with the same result by Michelson. *The theory of Hertz must therefore be rejected.*

II

The Principle of Relativity

But if the ether is not carried along by the motion of the earth, is it possible to show, by means of optical phenomena, the absolute velocity of the earth, or rather its velocity with respect to the unmoving ether? Experiment has answered negatively, and yet the experimental procedures have been varied in all possible ways. Whatever be the means employed there will never be disclosed anything but relative velocities; I mean the velocities of certain material bodies with reference to other material bodies. In fact, if the source of light and the apparatus of observation are on the earth and participate in its motion, the experimental results have always been the same, whatever be the orientation of the apparatus with reference to the orbital motion of the earth. If astronomic aberration happens, it is because the source, a star, is in motion with reference to the observer.

The hypotheses so far made perfectly account for this general result, *if we neglect very small quantities of the order of the square of the aberration*. The explanation rests upon the notion of *local time*, introduced by Lorentz, which I shall try to make clear. Suppose two observers, placed one at A, the other at B, and wishing to set their watches by means of optical signals. They agree that B shall send a signal to A when his watch marks an hour determined upon, and A is to put his watch to that hour the moment he sees the signal. If this alone were done, there would be a systematic error, because as the light takes a certain time t to go from B to A, A 's watch would be behind B 's the time t. This error is easily corrected. It suffices to cross the signals. A in turn must signal B, and, after this new adjustment, B 's watch will be behind A 's the time t. Then it will be sufficient to take the arithmetic mean of the two adjustments.

But this way of doing supposes that light takes the same time to go from A to B as to return from B to A. That is true if the observers are motionless; it is no longer so if they are carried along in a common translation, since then A, for example, will go to meet the light coming from B, while B will flee before the light coming from A. If therefore the observers are borne along in a common translation and if they do not suspect it, their adjustment will be defective; their watches will not indicate the same time; each will show the *local time* belonging to the point where it is.

The two observers will have no way of perceiving this, if the unmoving ether can transmit to them only luminous signals all of the same velocity, and if the other signals they might send are transmitted by media carried along with them in their translation. The phenomenon each observes will be too soon or too late; it would be seen at the same instant only if the translation did not exist; but as it will be observed with a watch that is wrong, this will not be perceived and the appearances will not be altered.

It results from this that the compensation is easy to explain so long as we neglect the square of the aberration, and for a long time the experiments were not sufficiently precise to warrant taking account of it. But the day came when Michelson imagined a much more delicate procedure: he made rays interfere which had traversed different courses, after being reflected by mirrors; each of the paths approximating a meter and the fringes of interference permitting the recognition of a fraction of a thousandth of a millimeter, the square of the aberration could no longer be neglected, and *yet the results were still negative*. Therefore the theory required to be completed, and it has been by the *Lorentz-Fitzgerald hypothesis*.

These two physicists suppose that all bodies carried along in a translation undergo a contraction in the sense of this translation, while their dimensions perpendicular to this translation remain unchanged. *This contraction is the same for all bodies*; moreover, it is very slight, about one two-hundred-millionth for a velocity such as that of the earth. Furthermore our measuring instruments could not disclose it, even if they were much more precise; our measuring rods in fact undergo the same contraction as the objects to be measured. If the meter exactly fits when applied to a body, if we point the body and consequently the meter in the sense of the motion of the earth, it will not cease to exactly fit in another orientation, and that although the body and the meter have changed in length as well as orientation, and precisely because the change is the same for one as for the other. But it is quite different if we measure a length, not now with a meter, but by the time taken by light to pass along it, and this is just what Michelson has done.

A body, spherical when at rest, will take thus the form of a flattened ellipsoid of revolution when in motion; but the observer will always think it spherical, since he himself has undergone an analogous deformation, as also all the objects serving as points of reference. On the contrary, the surfaces of the waves of light, remaining rigorously spherical, will seem to him elongated ellipsoids.

What happens then? Suppose an observer and a source of light carried along together in the translation: the wave surfaces emanating from the source will be spheres having as centers the successive positions of the source; the distance from this center to the actual position of the source will be proportional to the time elapsed after the emission, that is to say to the radius of the sphere. All these spheres are therefore homothetic one to the other, with relation to the actual position S of the source. But, for our observer, because of the contraction, all these spheres will seem elongated ellipsoids, and all these ellipsoids will moreover be homothetic, with reference to the point S; the excentricity of all these ellipsoids is the same and depends solely upon the velocity of the earth. *We shall so select the law of contraction that the point S may be at the focus of the meridian section of the ellipsoid.*

This time the compensation is *rigorous*, and this it is which explains Michelson's experiment.

I have said above that, according to the ordinary theories, observations of the astronomic aberration would give us the absolute velocity of the earth, if our instruments were a thousand times more precise. I must modify this statement. Yes, the observed angles would be modified by the effect of this absolute velocity, but the graduated circles we use to measure the angles would be deformed by the translation: they would become ellipses; thence would result an error in regard to the angle measured, and *this second error would exactly compensate the first.*

This Lorentz-Fitzgerald hypothesis seems at first very extraordinary; all we can say for the moment, in its favor, is that it is only the immediate translation of Michelson's experimental result, if we *define* lengths by the time taken by light to run along them.

However that may be, it is impossible to escape the impression that the principle of relativity is a general law of nature, that one will never be able by any imaginable means to show any but relative velocities, and I mean by that not only the velocities of bodies with reference to the ether, but the velocities of bodies with regard to one another. Too many different experiments have given concordant results for us not to feel tempted to attribute to this principle of relativity a value comparable to that, for example, of the principle of equivalence. In any case, it is proper to see to what consequences this way of looking at things would lead us and then to submit these consequences to the control of experiment.

III

The Principle of Reaction

Let us see what the principle of the equality of action and reaction becomes in the theory of Lorentz. Consider an electron A which for any cause begins to move; it produces a perturbation in the ether; at the end of a certain time, this perturbation reaches another electron B, which will be disturbed from its position of equilibrium. In these conditions there can not be equality between action and reaction, at least if we do not consider the ether, but only the electrons, *which alone are observable*, since our matter is made of electrons.

In fact it is the electron A which has disturbed the electron B; even in case the electron B should react upon A, this reaction could be equal to the action, but in no case simultaneous, since the electron B can begin to move only after a certain time, necessary for the propagation. Submitting the problem to a more exact calculation, we reach the following result: Suppose a Hertz discharger placed at the focus of a parabolic mirror to which it is mechanically attached; this discharger emits electromagnetic waves, and the mirror reflects all these waves in the same direction; the discharger therefore will radiate energy in a determinate direction. Well, the calculation shows that *the discharger recoils* like a cannon which has shot out a projectile. In the case of the cannon, the recoil is the natural result of the equality of action and reaction. The cannon recoils because the projectile upon which it has acted reacts upon it. But here it is no longer the same. What has been sent out is no longer a material projectile: it is energy, and energy has no mass: it has no counterpart. And, in place of a discharger, we could have considered just simply a lamp with a reflector concentrating its rays in a single direction.

It is true that, if the energy sent out from the discharger or from the lamp meets a material object, this object receives a mechanical push as if it had been hit by a real projectile, and this push will be equal to the recoil of the discharger and of the lamp, if no energy has been lost on the way and if the object absorbs the whole of the energy. Therefore one is tempted to say that there still is compensation between the action and the reaction. But this compensation, even should it be complete, is always belated. It never happens if the light, after leaving its source, wanders through interstellar spaces without ever meeting a material body; it is incomplete, if the body it strikes is not perfectly absorbent.

Are these mechanical actions too small to be measured, or are they accessible to experiment? These actions are nothing other than those due to the *Maxwell-Bartholi* pressures; Maxwell had predicted these pressures from calculations relative to electrostatics and magnetism; Bartholi reached the same result by thermodynamic considerations.

This is how the *tails of comets* are explained. Little particles detach themselves from the nucleus of the comet; they are struck by the light of the sun, which pushes them back as would a rain of projectiles coming from the sun. The mass of these particles is so little that this repulsion sweeps it away against the Newtonian attraction; so in moving away from the sun they form the tails.

The direct experimental verification was not easy to obtain. The first endeavor led to the construction of the *radiometer*. But this instrument *turns backward*, in the sense opposite to the theoretic sense, and the explanation of its rotation, since discovered, is wholly different. At last success came, by making the vacuum more complete, on the one hand, and on the other by not blackening one of the faces of the paddles and directing a pencil of luminous rays upon one of the faces. The radiometric effects and the other disturbing causes are eliminated by a series of painstaking precautions, and one obtains a deviation which is very minute, but which is, it would seem, in conformity with the theory.

The same effects of the Maxwell-Bartholi pressure are forecast likewise by the theory of Hertz of which we have before spoken, and by that of Lorentz. But there is a difference. Suppose that the energy, under the form of light, for example, proceeds from a luminous source to any body through a transparent medium. The Maxwell-Bartholi pressure will act, not alone upon the source at the departure, and on the body lit up at the arrival, but upon the matter of the transparent medium which it traverses. At the moment when the luminous wave reaches a new region of this medium, this pressure will push forward the matter there distributed and will put it back when the wave leaves this region. So that the recoil of the source has for counterpart the forward movement of the transparent matter which is in contact with this source; a little later, the recoil of this same matter has for counterpart the forward movement of the transparent matter which lies a little further on, and so on.

Only, is the compensation perfect? Is the action of the Maxwell-Bartholi pressure upon the matter of the transparent medium equal to its reaction upon the source, and that, whatever be this matter? Or is this action by so much the less as the medium is less refractive and more rarefied, becoming null in the void?

If we admit the theory of Hertz, who regards matter as mechanically bound to the ether, so that the ether may be entirely carried along by matter, it would be necessary to answer yes to the first question and no to the second.

There would then be perfect compensation, as required by the principle of the equality of action and reaction, even in the least refractive media, even in the air, even in the interplanetary void, where it would suffice to suppose a residue of matter, however subtile. If on the contrary we admit the theory of Lorentz, the compensation, always imperfect, is insensible in the air and becomes null in the void.

But we have seen above that Fizeau's experiment does not permit of our retaining the theory of Hertz; it is necessary therefore to adopt the theory of Lorentz, and consequently *to renounce the principle of reaction.*

IV

Consequences of the Principle of Relativity

We have seen above the reasons which impel us to regard the principle of relativity as a general law of nature. Let us see to what consequences this principle would lead, should it be regarded as finally demonstrated.

First, it obliges us to generalize the hypothesis of Lorentz and Fitzgerald on the contraction of all bodies in the sense of the translation. In particular, we must extend this hypothesis to the electrons themselves. Abraham considered these electrons as spherical and indeformable; it will be necessary for us to admit that these electrons, spherical when in repose, undergo the Lorentz contraction when in motion and take then the form of flattened ellipsoids.

This deformation of the electrons will influence their mechanical properties. In fact I have said that the displacement of these charged electrons is a veritable current of convection and that their apparent inertia is due to the self-induction of this current: exclusively as concerns the negative electrons; exclusively or not, we do not yet know, for the positive electrons. Well, the deformation of the electrons, a deformation which depends upon their velocity, will modify the distribution of the electricity upon their surface, consequently the intensity of the convection current they produce, consequently the laws according to which the self-induction of this current will vary as a function of the velocity.

At this price, the compensation will be perfect and will conform to the requirements of the principle of relativity, but only upon two conditions :

1° That the positive electrons have no real mass, but only a fictitious electromagnetic mass; or at least that their real mass, if it exists, is not constant and varies with the velocity according to the same laws as their fictitious mass;

2° That all forces are of electromagnetic origin, or at least that they vary with the velocity according to the same laws as the forces of electromagnetic origin.

It still is Lorentz who has made this remarkable synthesis; stop a moment and see what follows therefrom. First, there is no more matter, since the positive electrons no longer have real mass, or at least no constant real mass. The present principles of our mechanics, founded upon the constancy of mass, must therefore be modified. Again, an electromagnetic explanation must be sought of all the known forces, in particular of gravitation, or at least the law of gravitation must be so modified that this force is altered by velocity in the same way as the electromagnetic forces. We shall return to this point.

All that appears, at first sight, a little artificial. In particular, this deformation of electrons seems quite hypothetical. But the thing may be presented otherwise, so as to avoid putting this hypothesis of deformation at the foundation of the reasoning. Consider the electrons as material points and ask how their mass should vary as function of the velocity not to contravene the principle of relativity. Or, still better, ask what should be their acceleration under the influence of an electric or magnetic field, that this principle be not violated and that we come back to the ordinary laws when we suppose the velocity very slight. We shall find that the variations of this mass, or of these accelerations, must be *as if* the electron underwent the Lorentz deformation.

V
Kaufmann's Experiment

We have before us, then, two theories: one where the electrons are indeformable, this is that of Abraham; the other where they undergo the Lorentz deformation. In both cases, their mass increases with the velocity, becoming infinite when this velocity becomes equal to that of light; but the law of the variation is not the same. The method employed by Kaufmann to bring to light the law of variation of the mass seems therefore to give us an experimental means of deciding between the two theories.

Unhappily, his first experiments were not sufficiently precise for that; so he decided to repeat them with more precautions, and measuring with great care the intensity of the fields. Under their new form *they are in favor of the theory of Abraham.* Then the principle of relativity would not have the rigorous value we were tempted to attribute to it; there would no longer be reason for believing the positive electrons denuded of real mass like the negative electrons. However, before definitely adopting this conclusion, a little reflection is necessary. The question is of such importance that it is to be wished Kaufmann's experiment were repeated by another experimenter.[1] Unhappily, this experiment is very delicate and could be carried out successfully only by a physicist of the same ability as Kaufmann. All precautions have been properly taken and we hardly see what objection could be made.

There is one point however to which I wish to draw attention: that is to the measurement of the electrostatic field, a measurement upon which all depends. This field was produced between the two armatures of a condenser; and, between these armatures, there was to be made an extremely perfect vacuum, in order to obtain a complete isolation. Then the difference of potential of the two armatures was measured, and the field obtained by dividing this difference by the distance apart of the armatures. That supposes the field uniform; is this certain? Might there not be an abrupt fall of potential in the neighborhood of one of the armatures, of the negative armature, for example? There may be a difference of potential at the meeting of the metal and the vacuum, and it may be that this difference is not the same on the positive side and on the negative side; what would lead me to think so is the electric valve effects between mercury and vacuum. However slight the probability that it is so, it seems that it should be considered.

VI
The Principle of Inertia

In the new dynamics, the principle of inertia is still true, that is to say that an *isolated* electron will have a straight and uniform motion. At least this is generally assumed; however,Lindemann has made objections to this view; I do not wish to take part in this discussion, which I can not here expound because of its too difficult character. In any case, slight modifications to the theory would suffice to shelter it from Lindemann's objections.

We know that a body submerged in a fluid experiences, when in motion, considerable resistance, but this is because our fluids are viscous; in an ideal fluid, perfectly free from viscosity, the body would stir up behind it a liquid hill, a sort of wake; upon departure, a great effort would be necessary to put it in motion, since it would be necessary to move not only the body itself, but the liquid of its wake. But, the motion once acquired, it would perpetuate itself without resistance, since the body, in advancing, would simply carry with it the perturbation of the liquid, without the total vis viva of the liquid augmenting. Everything would happen therefore as if its inertia was augmented. An electron advancing in the ether would behave in the same way: around it, the ether would be stirred up, but this perturbation would accompany the body in its motion; so that, for an observer carried along with the electron, the electric and magnetic fields accompanying this electron would appear invariable, and would change only if the velocity of the electron varied. An effort would therefore be necessary to put the electron in motion, since it would be necessary to create the energy of these fields; on the contrary, once the movement acquired, no effort would be necessary to maintain it, since the created energy would only have to go along behind the electron as a wake. This energy, therefore, could only augment the inertia of the electron, as the agitation of the liquid augments that of the body submerged in a perfect fluid. And anyhow, the negative electrons at least have no other inertia except that.

In the hypothesis of Lorentz, the vis viva, which is only the energy of the ether, is not proportional to v^2. Doubtless if v is very slight, the vis viva is sensibly proportional to v^2, the quantity of motion sensibly proportional to v, the two masses sensibly constant and equal to each other. But *when the velocity tends toward the velocity of light, the vis viva, the quantity of motion and the two masses increase beyond all limit.*

In the hypothesis of Abraham, the expressions are a little more complicated; but what we have just said remains true in essentials.

So the mass, the quantity of motion, the vis viva become infinite when the velocity is equal to that of light.

Thence results that *no body can attain in any way a velocity beyond that of light.* And in fact, in proportion as its velocity increases, its mass increases, so that its inertia opposes to any new increase of velocity a greater and greater obstacle.

A question then suggests itself: let us admit the principle of relativity; an observer in motion would not have any means of perceiving his own motion. If therefore no body in its absolute motion can exceed the velocity of light, but may approach it as nearly as you choose, it should be the same concerning its relative motion with reference to our observer. And then we might be tempted to reason as follows: The observer may attain a velocity of 200,000 kilometers; the body in its relative motion with reference to the observer may attain the same velocity; its absolute velocity will then be 400,000 kilometers, which is impossible, since this is beyond the velocity of light. This is only a seeming, which vanishes when account is taken of how Lorentz evaluates local time.

VII
The Wave of Acceleration

When an electron is in motion, it produces a perturbation in the ether surrounding it; if its motion is straight and uniform, this perturbation reduces to the wake of which we have spoken in the preceding section. But it is no longer the same, if the motion be curvilinear or varied. The perturbation may then be regarded as the superposition of two others, to which Langevin has given the names *wave of velocity* and *wave of acceleration*. The wave of velocity is only the wave which happens in uniform motion.

As to the wave of acceleration, this is a perturbation altogether analogous to light waves, which starts from the electron at the instant when it undergoes an acceleration, and which is then propagated by successive spherical waves with the velocity of light. Whence follows: in a straight and uniform motion, the energy is wholly conserved; but, when there is an acceleration, there is loss of energy, which is dissipated under the form of luminous waves and goes out to infinity across the ether.

However, the effects of this wave of acceleration, in particular the corresponding loss of energy, are in most cases negligible, that is to say not only in ordinary mechanics and in the motions of the heavenly bodies, but even in the radium rays, where the velocity is very great without the acceleration being so. We may then confine ourselves to applying the laws of mechanics, putting the force equal to the product of acceleration by mass, this mass, however, varying with the velocity according to the laws explained above. We then say the motion is *quasi-stationary*.

It would not be the same in all cases where the acceleration is great, of which the chief are the following:

1° In incandescent gases certain electrons take an oscillatory motion of very high frequency; the displacements are very small, the velocities are finite, and the accelerations very great; energy is then communicated to the ether, and this is why these gases radiate light of the same period as the oscillations of the electron;

2° Inversely, when a gas receives light, these same electrons are put in swing with strong accelerations and they absorb light;

3° In the Hertz discharger, the electrons which circulate in the metallic mass undergo, at the instant of the discharge, an abrupt acceleration and take then an oscillatory motion of high frequency. Thence results that a part of the energy radiates under the form of Hertzian waves;

4° In an incandescent metal, the electrons enclosed in this metal are impelled with great velocity; upon reaching the surface of the metal, which they can not get through, they are reflected and thus undergo a considerable acceleration. This is why the metal emits light. The details of the laws of the emission of light by dark bodies are perfectly explained by this hypothesis;

5° Finally when the cathode rays strike the anticathode, the negative electrons constituting these rays, which are impelled with very great velocity, are abruptly arrested. Because of the acceleration they thus undergo, they produce undulations in the ether. This, according to certain physicists, is the origin of the Röntgen rays, which would only be light rays of very short wave-length.

CHAPTER III

The New Mechanics and Astronomy

I

Gravitation

Mass may be defined in two ways:

1° By the quotient of the force by the acceleration; this is the true definition of the mass, which measures the inertia of the body.

2° By the attraction the body exercises upon an exterior body, in virtue of Newton's law. We should therefore distinguish the mass coefficient of inertia and the mass coefficient of attraction. According to Newton's law, there is rigorous proportionality between these two coefficients. But that is demonstrated only for velocities to which the general principles of dynamics are applicable. Now, we have seen that the mass coefficient of inertia increases with the velocity; should we conclude that the mass coefficient of attraction increases likewise with the velocity and remains proportional to the coefficient of inertia, or, on the contrary, that this coefficient of attraction remains constant? This is a question we have no means of deciding.

On the other hand, if the coefficient of attraction depends upon the velocity, since the velocities of two bodies which mutually attract are not in general the same, how will this coefficient depend upon these two velocities ?

Upon this subject we can only make hypotheses, but we are naturally led to investigate which of these hypotheses would be compatible with the principle of relativity. There are a great number of them; the only one of which I shall here speak is that of Lorentz, which I shall briefly expound.

Consider first electrons at rest. Two electrons of the same sign repel each other and two electrons of contrary sign attract each other; in the ordinary theory, their mutual actions are proportional to their electric charges; if therefore we have four electrons, two positive A and A', and two negative B and B', the charges of these four being the same in absolute value, the repulsion of A for A' will be, at the same distance, equal to the repulsion of B for B' and equal also to the attraction of A for B', or of A' for B. If therefore A and B are very near each other, as also A' and B', and we examine the action of the system $A + B$ upon the system $A' + B'$, we shall have two repulsions and two attractions which will exactly compensate each other and the resulting action will be null.

Now, material molecules should just be regarded as species of solar systems where circulate the electrons, some positive, some negative, and *in such a way that the algebraic sum of all the charges is null*. A material molecule is therefore wholly analogous to the system $A + B$ of which we have spoken, so that the total electric action of two molecules one upon the other should be null.

But experiment shows us that these molecules attract each other in consequence of Newtonian gravitation; and then we may make two hypotheses: we may suppose gravitation has no relation to the electrostatic attractions, that it is due to a cause entirely different, and is simply something additional; or else we may suppose the attractions are not proportional to the charges and that the attraction exercised by a charge $+1$ upon a charge -1 is greater than the mutual repulsion of two $+1$

charges, or two —1 charges.

In other words, the electric field produced by the positive electrons and that which the negative electrons produce might be superposed and yet remain distinct. The positive electrons would be more sensitive to the field produced by the negative electrons than to the field produced by the positive electrons; the contrary would be the case for the negative electrons. It is clear that this hypothesis somewhat complicates electrostatics, but that it brings back into it gravitation. This was, in sum, Franklin's hypothesis.

What happens now if the electrons are in motion? The positive electrons will cause a perturbation in the ether and produce there an electric and a magnetic field. The same will be the case for the negative electrons. The electrons, positive as well as negative, undergo then a mechanical impulsion by the action of these different fields. In the ordinary theory, the electromagnetic field, due to the motion of the positive electrons, exercises, upon two electrons of contrary sign and of the same absolute charge, equal actions with contrary sign. We may then without inconvenience not distinguish the field due to the motion of the positive electrons and the field due to the motion of the negative electrons and consider only the algebraic sum of these two fields, that is to say the resulting field.

In the new theory, on the contrary, the action upon the positive electrons of the electromagnetic field due to the positive electrons follows the ordinary laws; it is the same with the action upon the negative electrons of the field due to the negative electrons. Let us now consider the action of the field due to the positive electrons upon the negative electrons (or inversely); it will still follow the same laws, but *with a different coefficient*. Each electron is more sensitive to the field created by the electrons of contrary name than to the field created by the electrons of the same name.

Such is the hypothesis of Lorentz, which reduces to Franklin's hypothesis for slight velocities; it will therefore explain, for these small velocities, Newton's law. Moreover, as gravitation goes back to forces of electrodynamic origin, the general theory of Lorentz will apply, and consequently the principle of relativity will not be violated.

We see that Newton's law is no longer applicable to great velocities and that it must be modified, for bodies in motion, precisely in the same way as the laws of electrostatics for electricity in motion.

We know that electromagnetic perturbations spread with the velocity of light. We may therefore be tempted to reject the preceding theory upon remembering that gravitation spreads, according to the calculations of Laplace, at least ten million times more quickly than light, and that consequently it can not be of electromagnetic origin. The result of Laplace is well known, but one is generally ignorant of its signification. Laplace supposed that, if the propagation of gravitation is not instantaneous, its velocity of spread combines with that of the body attracted, as happens for light in the phenomenon of astronomic aberration, so that the effective force is not directed along the straight joining the two bodies, but makes with this straight a small angle. This is a very special hypothesis, not well justified, and, in any case, entirely different from that of Lorentz. Laplace's result proves nothing against the theory of Lorentz.

II

Comparison with Astronomic Observations

Can the preceding theories be reconciled with astronomic observations?

First of all, if we adopt them, the energy of the planetary motions will be constantly dissipated by the effect of the *wave of acceleration*. From this would result that the mean motions of the stars would constantly accelerate, as if these stars were moving in a resistant medium. But this effect is exceedingly slight, far too much so to be discerned by the most precise observations. The acceleration of the heavenly bodies is relatively slight, so that the effects of the wave of acceleration are negligible and the motion may be regarded as *quasi-stationary*. It is true that the effects of the wave of acceleration constantly accumulate, but this accumulation itself is so slow that thousands of years of observation would be necessary for it to become sensible. Let us therefore make the calculation considering the motion as quasi-stationary, and that under the three following hypotheses :

A. Admit the hypothesis of Abraham (electrons indeformable) and retain Newton's law in its usual form;

B. Admit the hypothesis of Lorentz about the deformation of electrons and retain the usual Newton's law;

C. Admit the hypothesis of Lorentz about electrons and modify Newton's law as we have done in the preceding paragraph, so as to render it compatible with the principle of relativity.

It is in the motion of Mercury that the effect will be most sensible, since this planet has the greatest velocity. Tisserand formerly made an analogous calculation, admitting Weber's law; I recall that Weber had sought to explain at the same time the electrostatic and electrodynamic phenomena in supposing that electrons (whose name was not jet invented) exercise, one upon another, attractions and repulsions directed along the straight joining them, and depending not only upon their distances, but upon the first and second derivatives of these distances, consequently upon their velocities and their accelerations. This law of Weber, different enough from those which to-day tend to prevail, none the less presents a certain analogy with them.

Tisserand found that, if the Newtonian attraction conformed to Weber's law there resulted, for Mercury's perihelion, secular variation of 14", *of the same sense as that which has been observed and could not be explained*, but smaller, since this is 38".

Let us recur to the hypotheses A, B and C, and study first the motion of a planet attracted by a fixed center. The hypotheses B and C are no longer distinguished, since, if the attracting point is fixed, the field it produces is a purely electrostatic field, where the attraction varies inversely as the square of the distance, in conformity with Coulomb's electrostatic law, identical with that of Newton.

The vis viva equation holds good, taking for vis viva the new definition; in the same way, the equation of areas is replaced by another equivalent to it; the moment of the quantity of motion is a constant, but the quantity of motion must be defined as in the new dynamics.

The only sensible effect will be a secular motion of the perihelion. With the theory of Lorentz, we shall find, for this motion, half of what Weber's law would give; with the theory of Abraham, two fifths.

If now we suppose two moving bodies gravitating around their common center of gravity, the effects are very little different, though the calculations may be a little more complicated. The motion of Mercury's perihelion would therefore be 7" in the theory of Lorentz and 5".6 in that of Abraham.

The effect moreover is proportional to n^2a^2, where n is the star's mean motion and a the radius of its orbit. For the planets, in virtue of Kepler's law, the effect varies then inversely as $\sqrt{a^5}$; it is therefore insensible, save for Mercury.

It is likewise insensible for the moon though n is great, because a is extremely small; in sum, it is five times less for Venus, and six hundred times less for the moon than for Mercury. We may add that as to Venus and the earth, the motion of the perihelion (for the same angular velocity of this motion) would be much more difficult to discern by astronomic observations, because the excentricity of their orbits is much less than for Mercury.

To sum up, *the only sensible effect upon astronomic observations would be a motion of Mercury's perihelion, in the same sense as that which has been observed without being explained, but notably slighter.*

That can not be regarded as an argument in favor of the new dynamics, since it will always be necessary to seek another explanation for the greater part of Mercury's anomaly; but still less can it be regarded as an argument against it.

III

The Theory of Lesage

It is interesting to compare these considerations with a theory long since proposed to explain universal gravitation. .

Suppose that, in the interplanetary spaces, circulate in every direction, with high velocities, very tenuous corpuscles. A body isolated in space will not be affected, apparently, by the impacts of these corpuscles, since these impacts are equally distributed in all directions. But if two bodies A and B are present, the body B will play the role of screen and will intercept part of the corpuscles which, without it, would have struck A. Then, the impacts received by A in the direction opposite that from B will no longer have a counterpart, or will now be only partially compensated, and this will push A toward B.

Such is the theory of Lesage; and we shall discuss it, taking first the view-point of ordinary mechanics.

First, how should the impacts postulated by this theory take place; is it according to the laws of perfectly elastic bodies, or according to those of bodies devoid of elasticity, or according to an intermediate law? The corpuscles of Lesage can not act as perfectly elastic bodies; otherwise the effect would be null, since the corpuscles intercepted by the body B would be replaced by others which would have rebounded from B, and calculation proves that the compensation would be perfect. It is necessary then that the impact make the corpuscles lose energy, and this energy should appear under the form of heat. But how much heat would thus be produced? Note that attraction passes through bodies; it is necessary therefore to represent to ourselves the earth, for example, not as a solid screen, but as formed of a very great number of very small spherical molecules, which play individually the rôle of little screens, but between which the corpuscles of Lesage may freely circulate. So, not only the earth is not a solid screen, but it is not even a cullender, since the voids occupy much more space than the plenums. To realize this, recall that Laplace has demonstrated that attraction, in traversing the earth, is weakened at most by one ten-millionth part, and his proof is perfectly satisfactory: in fact, if attraction were absorbed by the body it traverses, it would no longer be proportional to the masses; it would be *relatively* weaker for great bodies than for small, since it would have a greater thickness to traverse. The attraction of the sun for the earth would therefore be *relatively* weaker than that of the sun for the moon, and thence would result, in the motion of the moon, a very sensible inequality. We should therefore conclude, if we adopt the theory of Lesage, that the total surface of the spherical molecules which compose the earth is at most the ten-millionth part of the total surface of the earth.

Darwin has proved that the theory of Lesage only leads exactly to Newton's law when we postulate particles entirely devoid of elasticity. The attraction exerted by the earth on a mass 1 at a distance 1 will then be proportional, at the same time, to the total surface S of the spherical molecules composing it, to the velocity v of the corpuscles, to the square root of the density ρ of the medium formed by the corpuscles. The heat produced will be proportional to S, to the density ρ, and to the cube of the velocity v.

But it is necessary to take account of the resistance experienced by a body moving in such a medium; it can not move, in fact, without going against certain impacts, in fleeing, on the contrary, before those coming in the opposite direction, so that the compensation realized in the state of rest can no longer subsist. The calculated resistance is proportional to S, to ρ and to v; now, we know that the heavenly bodies move as if they experienced no resistance, and the precision of observations permits us to fix a limit to the resistance of the medium.

This resistance varying as $S\rho v$, while the attraction varies as $S\sqrt{\rho v}$, we see that the ratio of the resistance to the square of the attraction is inversely as the product Sv.

We have therefore a lower limit of the product Sv. We have already an upper limit of S (by the absorption of attraction by the body it traverses); we have therefore a lower limit of the velocity v, which must be at least $24 \cdot 10^{17}$ times that of light.

From this we are able to deduce ρ and the quantity of heat produced; this quantity would suffice to raise the temperature 10^{26} degrees a second; the earth would receive in a given time 10^{20} times more heat than the sun emits in the same time; I am not speaking of the heat the sun sends to the earth, but of that it radiates in all directions.

It is evident the earth could not long stand such a regime.

We should not be led to results less fantastic if, contrary to Darwin's views, we endowed the corpuscles of Lesage with an elasticity imperfect without being null. In truth, the vis viva of these corpuscles would not be entirely converted into heat, but the attraction produced would likewise be less, so that it would be only the part of this vis viva converted into heat, which would contribute to produce the attraction and that would come to the same thing; a judicious employment of the theorem of the viriel would enable us to account for this.

The theory of Lesage may be transformed; suppress the corpuscles and imagine the ether overrun in all senses by luminous waves coming from all points of space. When a material object receives a luminous wave, this wave exercises upon it a mechanical action due to the Maxwell-Bartholi pressure, just as if it had received the impact of a material projectile. The waves in question could therefore play the role of the corpuscles of Lesage. This is what is supposed, for example, by M. Tommasina.

The difficulties are not removed for all that; the velocity of propagation can be only that of light, and we are thus led, for the resistance of the medium, to an inadmissible figure. Besides, if the light is all reflected, the effect is null, just as in the hypothesis of the perfectly elastic corpuscles.

That there should be attraction, it is necessary that the light be partially absorbed; but then there is production of heat. The calculations do not differ essentially from those made in the ordinary theory of Lesage, and the result retains the same fantastic character.

On the other hand, attraction is not absorbed by the body it traverses, or hardly at all; it is not so with the light we know. Light which would produce the Newtonian attraction would have to be considerably different from ordinary light and be, for example, of very short wave length. This does not count that, if our eyes were sensible of this light, the whole heavens should appear to us much more brilliant than the sun, so that the son would seem to us to stand out in black, otherwise the sun would repel us instead of attracting us. For all these reasons, light which would permit of the explanation of attraction would be much more like Röntgen rays than like ordinary light.

And besides, the X-rays would not suffice; however penetrating they may seem to us, they could not pass through the whole earth; it would be necessary therefore to imagine X'-rays much more penetrating than the ordinary X-rays. Moreover a part of the energy of these X'-rays would have to be destroyed, otherwise there would be no attraction. If you do not wish it transformed into heat, which would lead to an enormous heat production, you must suppose it radiated in every direction under the form of secondary rays, which might be called X'' and which would have to be much more penetrating still than the X'-rays, otherwise they would in their turn derange the phenomena of attraction.

Such are the complicated hypotheses to which we are led when we try to give life to the theory of Lesage.

But all we have said presupposes the ordinary laws of mechanics.

Will things go better if we admit the new dynamics? And first, can we conserve the principles of relativity? Let us give at first to the theory of Lesage its primitive form, and suppose space ploughed by material corpuscles; if these corpuscles were perfectly elastic, the laws of their impact would conform to this principle of relativity, but we know that then their effect would be null. We must therefore suppose these corpuscles are not elastic, and then it is difficult to imagine a law of impact compatible with the principle of relativity. Besides, we should still find a production of considerable heat, and yet a very sensible resistance of the medium.

If we suppress these corpuscles and revert to the hypothesis of the Maxwell-Bartholi pressure, the difficulties will not be less. This is what Lorentz himself has attempted in his Memoir to the Amsterdam Academy of Sciences of April 25, 1900.

Consider a system of electrons immersed in an ether permeated in every sense by luminous waves; one of these electrons, struck by one of these waves, begins to vibrate; its vibration will be synchronous with that of light; but it may have a difference of phase, if the electron absorbs a part of the incident energy. In fact, if it absorbs energy, this is because the vibration of the ether *impels* the electron; the electron must therefore be slower than the ether. An electron in motion is analogous to a convection current; therefore every magnetic field, in particular that due to the luminous perturbation itself, must exert a mechanical action upon this electron. This action is very slight; moreover, it changes sign in the current of the period; nevertheless, the mean action is not null if there is a difference of phase between the vibrations of the electron and those of the ether. The mean action is proportional to this difference, consequently to the energy absorbed by the electron. I can not here enter into the detail of the calculations; suffice it to say only that the final result is an attraction of any two electrons, varying inversely as the square of the distance and proportional to the energy absorbed by the two electrons.

Therefore there can not be attraction without absorption of light and, consequently, without production of heat, and this it is which determined Lorentz to abandon this theory, which, at bottom, does not differ from that of Lesage-Maxwell-Bartholi. He would have been much more dismayed still if he had pushed the calculation to the end. He would have found that the temperature of the earth would have to increase 10^{13} degrees a second.

IV

Conclusions

I have striven to give in few words an idea as complete as possible of these new doctrines; I have sought to explain how they took birth; otherwise the reader would have had ground to be frightened by their boldness. The new theories are not yet demonstrated; far from it; only they rest upon an aggregate of probabilities sufficiently weighty for us not to have the right to treat them with disregard.

New experiments will doubtless teach us what we should finally think of them. The knotty point of the question lies in Kaufmann's experiment and those that may be undertaken to verify it.

In conclusion, permit me a word of warning. Suppose that, after some years, these theories undergo new tests and triumph; then our secondary education will incur a great danger: certain professors will doubtless wish to make a place for the new theories.

Novelties are so attractive, and it is so hard not to seem highly advanced! At least there will be the wish to open vistas to the pupils and, before teaching them the ordinary mechanics, to let them know it has had its day and was at best good enough for that old dolt Laplace. And then they will not form the habit of the ordinary mechanics.

Is it well to let them know this is only approximative? Yes; but later, when it has penetrated to their very marrow, when they shall have taken the bent of thinking only through it, when there shall no longer be risk of their unlearning it, then one may, without inconvenience, show them its limits.

It is with the ordinary mechanics that they must live; this alone will they ever have to apply. Whatever be the progress of automobilism, our vehicles will never attain speeds where it is not true. The other is only a luxury, and we should think of the luxury only when there is no longer any risk of harming the necessary.

1.   At the moment of going to press we learn that M. Bucherer has repeated the experiment, taking new precautions, and that he has obtained, contrary to Kaufmann, results confirming the views of Lorentz.

The New Mechanics (1913)
by Henri Poincaré, translated by George Bruce Halsted

THE NEW MECHANICS.[1]

IN this world, as you know, nothing is final, nothing immutable; the most powerful, the most stable empires are not eternal: this is a theme the preachers abundantly develop.

Scientific theories are like empires, they are not certain of the morrow. If any one of them seemed beyond the effects of time, it was certainly the Newtonian mechanics. It seemed undisputed, it was an imperishable monument; and behold in its turn, I shall not say the monument is thrown down, that would be premature, but anyhow it is greatly shaken. It is subjected to the attacks of powerful destroyers. There is one in Göttingen, Max Abraham, another is the Dutch physicist Lorentz. I wish to say a few words about the ruins of the ancient edifice and about the new structure by which it is sought to replace them.

First of all what is it that characterizes the old mechanics? It is this very simple fact: Consider a body at rest, impart to it an impulse, that is to say make a given force act upon it for a given time; the body moves, acquires a certain velocity; the body being impelled by this velocity, apply to it again the same force for the same time, the velocity will be doubled; if we still continue, the velocity will be tripled when we shall have given a third time the same impulse. So beginning again a sufficient number of times, the body will end by acquiring a very great velocity which can exceed all limit, an infinite velocity.

In the new mechanics, on the contrary, we assume it to be impossible to communicate to a body starting from rest a velocity beyond that of light. What happens? Consider the same body at rest; give it a first impulse, the same as before, it will take the same velocity; repeat this impulse a second time, the velocity again augments, but it no longer will be doubled; a third impulse will produce an analogous effect, the velocity increases but less and less, the body opposing a resistance which becomes greater and greater. This resistance is inertia, it is what is commonly called mass; all this happens then in this new mechanics as if the mass was not constant, but increased with the velocity.

We can represent the phenomenon graphically: In the old mechanics the body after the first impulse takes a velocity represented by the sect On_1; after the second impulse On_1 increases by a sect $n_1 n_2$ equal to it; at each new

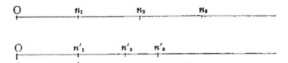

impulse the velocity increases by the same quantity, the sect representing it increasing by a constant length. In the new mechanics, the velocity sect increases by sects $n'_1 n'_1$, $n'_2 n'_3$, $\cdots$ which become smaller and smaller so that we cannot pass beyond a certain limit, the velocity of light.

How have we been led to such conclusions? Have we made direct experiments? The divergences only come out for bodies impelled by great velocities; only then do the indicated differences become perceptible. But what is a very great velocity? Is it that of an automobile making 100 kilometers an hour? We would go wild with excitement over such a speed in the street. But from our point of view this velocity is still very small, a snail's pace. Astronomy does better. Mercury, the fastest of the heavenly bodies, also runs over about 100 kilometers, no longer per hour but per second. However, that still does not suffice; such velocities are too slight to reveal the differences we wish to observe. I do not mention our cannon balls; they are faster than automobiles, but much slower than Mercury.

You know that we have discovered an artillery whose projectiles are much swifter; I mean radium which sends out energy projectiles in every direction. The speed of this shooting is far greater, the initial velocity being about 100,000 kilometers a second, one-third the velocity of light. The caliber of the projectiles and their weight are, it is true much slighter and we cannot count on this artillery to increase the fighting power of our armies.

Can we experiment on these projectiles? Such experiments have been actually undertaken; under the influence of an electric field, of a magnetic field, a deviation occurs which enables us to take account of the inertia and to measure it. Thus we have ascertained that the mass depends upon the velocity and we enunciate this law: The inertia of a body increases with its velocity which remains less than that of light, 300,000 kilometers a second.

I pass now to the second principle, the principle of relativity. Suppose there is an observer moving to the right; everything is as if he were at rest, with the objects about him moving to the left. There is no way of knowing whether the objects really move, whether the observer is at rest or in motion. We teach in all courses on mechanics that the passenger on a boat thinks he sees the river bank moving, while he is gently borne along by the motion of the boat. Examined more closely, this simple idea acquires capital importance; there is no way of settling the question, no experiment can disprove the principle that there is no absolute space, all displacements we can observe are relative displacements. I have often had occasion to express these considerations so familiar to philosophers. They have even given me a publicity I would gladly have avoided. All the reactionary French journals have made me prove that the sun turns around the earth. In the famous case between the Inquisition and Galileo, Galileo should be all wrong.

To return to the old mechanics. It admitted the principle of relativity; in place of being founded on experiments, its laws were deduced from this fundamental principle. These considerations sufficed for purely mechanical phenomena, but not for important parts of physics, for example optics. We considered the velocity of light as absolute with reference to the ether. This velocity could be measured. We had theoretically the means of comparing the displacement of a moving body to an absolute displacement, the means of deciding whether or not a body was in absolute motion.

Delicate experiments, apparatus exceedingly precise, which I shall not describe to you, enabled us to attempt the practical realization of such a comparison: the result was null. The principle of relativity admits of no restriction in the new mechanics; it has, if I may so speak, an absolute value.

To understand the role the principle of relativity plays in the new mechanics, we are led first to speak of apparent time, a very ingenious invention of the physicist Lorentz. We suppose two observers, the one A at Paris, the other B at Berlin. A and B have identical chronometers and wish to set them; but they are exceptionally scrupulous observers and require in their setting an extraordinary exactitude not only, for instance, to the second, but to the thousand-millionth of a second. How can they do it? From Paris to Berlin, A sends a telegraphic signal, by wireless, if you will, to be wholly modern. B notes the moment of reception and this will be the starting time for both chronometers. But it takes a certain time for the signal to go from Paris to Berlin; it travels only with the speed of light. B's watch would therefore be slow. B is too intelligent not to take this into account, and he proceeds to remedy it. The thing seems very simple. They cross signals, A receiving and B sending; they take the mean of the corrections thus made and so have the exact time.

But is this certain? We are assuming that it takes the signal the same time to go from A to B as from B to A. Now A and B are carried along in the motion of the earth with reference to the ether, the vehicle of the electric waves. When A has sent his signal it flies on before him, B moving away in the same way, and the time employed will be longer than if the two observers were at rest. If, on the other hand, it is B who sends, and A who receives, the time is shorter because A goes to meet the signal. It is absolutely impossible for them to know whether or not their chronometers mark the same time. Whatever the method employed, the troubles remain the same. The observation of an astronomic phenomenon and all optical methods run against the same difficulties. B can never know more than an apparent difference of time, more than a species of local hour. The principle of relativity applies completely.

In the old mechanics, however, we prove with this principle all the fundamental laws. We might be tempted to take up the classic arguments and reason as follows. Suppose again two observers, A and B, to call them what we always call two observers in mathematics. Suppose them in motion, going away from each other. Neither can surpass the velocity of light; for example let B go at the rate of 200,000 kilometers toward the right, A of 200,000 kilometers toward the left. A may think himself at rest, and the apparent velocity of B will for him be 400,000 kilometers. If A knows the new mechanics he will say: "B has a velocity he cannot attain, therefore I also am in motion." It seems he could decide about his absolute state. But he must be able to observe the motion of B himself. To make this observation A and B commence by setting their watches, then B sends telegrams to A to indicate to him his successive positions; putting them together A can reckon B's motion and trace the curve of this motion. Now the signals go with the speed of light. The watches which mark the apparent time vary at each instant and everything will happen as if B's watch went too fast. B will think himself going much less rapidly and the apparent velocity he will have relatively to A will not surpass the limit it should not attain. Nothing can reveal to A whether he is in motion or at absolute rest.

It is still necessary to make a third hypothesis, which is much more surprising, much more difficult to admit, and which greatly disturbs our present modes of thought. A body in motion of translation undergoes a deformation in the direction of its displacement; a sphere, for instance, becomes like a species of flattened ellipsoid with the short axis parallel to the translation. If we do not perceive such a transformation every day this is because it is so small as to render it almost imperceptible. The earth, borne along in its revolution through its orbit, is deformed about $1/200\,000\,000$. To observe such a phenomenon would require measuring instruments of extreme precision, but if their precision were infinite it would not avail, because they also are borne along in the motion and would undergo the same transformation. We should perceive nothing; the meter we could use would shorten like the length to be measured. We could not learn anything except by comparing the length of one of these bodies to the velocity of light.

These are delicate experiments, carried out by **Michelson** and I shall not expound their details ; they have given results altogether remarkable. However strange it may seem to us, it is necessary to admit that the third hypothesis is perfectly verified.

Such are the bases of the new mechanics; with the help of these hypotheses we find that it is compatible with the principle of relativity.

But it is necessary to connect it then to a new conception of matter.

For the modern physicist, the atom is no longer the simple element; it has become a veritable universe in which thousands of planets gravitate around tiny suns. Suns and planets are here particles *electrified* either negatively or positively; the physicist calls them *electrons* and with them builds the world. Some represent the neutral atom as a positive central mass around which circulate a great number of negatively charged electrons, whose total electric mass equals in magnitude that of the central nucleus.

This conception of matter enables us easily to account for the augmentation of the mass of a body with its velocity, which we have made one of the characteristics of the new mechanics. Since a body is only an assemblage of electrons, it will suffice to show it to be true of them. To this end we note that an isolated electron moving through the ether engenders an electric current, that is to say an electromagnetic field. This field corresponds to a certain quantity of energy localized not in the electron but in the ether. A variation in magnitude or in direction of the electron's velocity modifies the field and expresses itself by a variation of the electromagnetic energy of the ether. While in the Newtonian mechanics the expenditure of energy is due only to the inertia of the moving body, here a part of this expenditure is due to what may be called the inertia of the ether relatively to the electromagnetic forces. The inertia of the ether increases with the velocity and its limit becomes infinite when the velocity approaches the velocity of light. The apparent mass of the electron therefore increases with the velocity; Kaufmann's experiments show that the constant real mass of the electron is negligible in relation to the apparent mass and may be considered as null.

In the new conception, matter's constant mass has disappeared. Only the ether, and no longer matter, is inert. Only the ether opposes a resistance to motion, so that we might say that there is no matter, but only gaps in the ether. For stationary or quasi-stationary motions, the new mechanics does not differ — within the range of approximation of our measurements — from the Newtonian mechanics, with the sole difference that the mass is no longer independent either of the velocity or of the angle this velocity makes with the direction of the accelerative force. If, *per contra*, the velocity has a considerable acceleration, in the case, for instance, of very rapid oscillations, Hertzian waves are produced which represent a loss of energy of the electron involving the deadening of its motion. Thus in wireless telegraphy the waves emitted are due to the vibrations of the electrons in the oscillatory discharge.

Analogous vibrations take place in a flame and likewise also in an incandescent solid. Lorentz thinks that in an incandescent body a considerable number of electrons circulate which, not being able to get out of it, fly in every direction and are reflected on its surface. We may compare them to a swarm of gnats enclosed in a jar and striking with their wings against the walls of their prison. The higher the temperature, the more rapid becomes the motion of these electrons and the more numerous the mutual impacts and the reflections on the wall. At each impact and at each reflection an electromagnetic wave is emitted and it is the perception of these waves which makes the body appear to us incandescent.

The motion of the electrons is almost tangible in a Crookes tube. There a veritable bombardment takes place of electrons issuing from the cathode. These cathode rays violently strike the anticathode and are there in part reflected, thus giving birth to an electromagnetic agitation which many physicists identify with the Röntgen rays.

In closing, it remains for us to examine the relations of the new mechanics to astronomy.

If the notion of constant mass of a body vanishes, what will become of Newton's law? It will hold good only for bodies at rest. Moreover it will be necessary to take into account the fact that attraction is not instantaneous. It may therefore well be asked whether the new mechanics will not result in complicating astronomy without obtaining an approximation superior to that given by the classic celestial mechanics. Lorentz has taken up the question. Starting from Newton's law, which he assumes to be true for two electrified bodies at rest, he calculates the electrodynamic action of the currents engendered by these bodies in motion. He thus obtains a new law of attraction containing the velocities of the two bodies as parameters.

Before examining how this law explains astronomic phenomena, we remark again that the acceleration of the heavenly bodies has as consequence an electromagnetic radiation, therefore a dissipation of energy making itself felt in return by a deadening of their velocity. Therefore, in the long run, the planets will end by falling into the sun. But this prospect can hardly frighten us, since the catastrophe can not happen for some millions of milliards of centuries.

Returning now to the law of attraction, we easily see that the difference between the two mechanics will be the greater the greater the velocity of the planets.

If there is an appreciable difference, it will therefore be greatest for **Mercury**, which has the greatest velocity of all the planets. Now it happens precisely that Mercury presents an anomaly not yet explained. The motion of its perihelion is more rapid than the motion calculated by the classic theory. The acceleration is 38" too great. **Leverrier** attributed this anomaly to a planet not yet discovered and an amateur astronomer thought he observed its passage across the sun. Since then no one else has seen it and it is unhappily certain that this planet perceived was only a bird.

Now the new mechanics explains perfectly the sense of the error with regard to Mercury, but it still leaves a margin of 32" between it and observation. It therefore does not suffice for bringing concord into the explanation of the velocity of Mercury. If this result is hardly decisive in favor of the new mechanics, still less is it unfavorable to its acceptance since the sense in which it corrects the deviation from the classic theory is the right one. Our explanation of the velocity of the other planets is not sensibly modified in the new theory and the results coincide, to within the approximation of the measurements, with those of the classic theory.

In conclusion, it would be premature, I believe, in spite of the great value of the arguments and of the facts set up against it, to regard the classic mechanics as finally condemned. However it may be in other respects, it will remain the mechanics of very small velocities in relation to that of light, the mechanics therefore of our practical life and of our terrestrial technic. If however, in some years, its rival triumphs, I shall venture to point out a pedagogic danger that a number of teachers, in France at least, will not escape. These teachers will find nothing more important, in teaching elementary mechanics to their scholars, than to inform them that this mechanics has had its day, that a new mechanics where the notions of mass and of time have a wholly different value replaces it; they will look down upon this lapsed mechanics that the programs force them to teach and will make their scholars feel the contempt they have for it. Yet I believe that this disdained classic mechanics will be as necessary as now and that whoever does not know it thoroughly cannot understand the new mechanics.

HENRI POINCARÉ.

---

1. Translated from the French by **George Bruce Halsted**.

225

Made in the USA
Middletown, DE
21 August 2022

71905129R00126